Numerical Analysis

The techniques of Numerical Analysis have become of great importance in the development of mathematical modelling with the aid of computers. This book discusses a wide range of techniques, suggests their advantages and disadvantages, and accompanies the majority of them with some form of error analysis. Throughout, the mathematics has been kept as simple as is compatible with demonstrating a number of important results and properties of the various techniques.

DR. ALEXANDER GRAHAM studied Engineering and Mathematics in Paris and Dublin. He became a Research Fellow sponsored by Standard Telephones and Cables, and then a Lecturer in Mathematics at Woolwich Polytechnic in London. Subsequently he held a joint appointment as Lecturer in Numerical Analysis at Bedford College and at the Institute of Computer Science, University of London. In 1969 he was appointed Senior Lecturer in Mathematics at the Open University.

Transworld Student Library is a series of books devoted to topics in Mathematics and the Sciences designed to meet the modern educational needs of the independent learner. The academic level varies with the familiarity of the material covered, from general introductory presentations suitable for sixth-form pupils and school leavers to more specialized topics treated at first- or second-year University level. At the same time, many of the books will be of interest to the general reader with an enquiring mind who wishes to become acquainted with some of the more recent developments in Mathematics and the Sciences.

For the most part, the treatment breaks away from the traditional presentation geared to the classroom situation and provides a refreshingly new approach to the subject matter being discussed.

Transworld Student Library

General Editor

H. GRAHAM FLEGG, M.A., D.C.Ae., C.Eng., F.I.M.A.
M.I.E.E., A.F.R.Ae.S., F.R.Met.S.
Reader in Mathematics, The Open University

Other books in this series

Boolean Algebra   H. G. Flegg
Theoretical Statistics   S. N. Collings
Points and Arrows: the Theory of Graphs   A. Kaufmann
Meteorology   H-J. Tanck
Field Projects in Sociology   J. P. Wiseman and M. S. Aron
The Unknown Ego   T. Brocher
Calculus via Numerical Analysis   A. Graham and G. Read
Basic Mathematical Structures I   N. Gowar and H. G. Flegg
Towards Quantum Mechanics   J. Cunningham
Reliability: a Mathematical Approach   A. Kaufmann
Paradoxes of Physics   P. Chambadal
New Perspectives in the Theory of Evolution   H. Querner,
   H. Holder, A. Egelhaaf, J. Jacobs and G. Heberer
Numerical Analysis   A. Graham
Biology: the Ultimate Science   B. Hocking

NUMERICAL ANALYSIS

Alexander Graham

Senior Lecturer in Mathematics
The Open University

TRANSWORLD PUBLISHERS LTD
A National General Company
In association with Richard Sadler Ltd

NUMERICAL ANALYSIS

A TRANSWORLD STUDENT LIBRARY BOOK 0 552 40013 0

First publication in Great Britain

PRINTING HISTORY
Transworld Student Library Edition published 1973
in association with Richard Sadler Ltd

Copyright © 1973 by Alexander Graham

Transworld Student Library Books are published by
Transworld Publishers Ltd, Cavendish House,
57–59 Uxbridge Road, Ealing, London, W.5

Set in Times 10/11 pt

Made and printed in Great Britain by
Richard Clay (The Chaucer Press) Ltd
Bungay, Suffolk

**NOTE: The Australian price appearing on the
back cover is the recommended retail price**

CONTENTS

INTRODUCTION

Only a limited number of topics arising in Numerical Analysis are discussed in this small volume. This book has been written using as simple mathematics as is compatible with demonstrating a number of important results and properties of the various techniques discussed. Although when possible each technique is derived from 'first principles', it was not possible in the space available to discuss in detail each step taken and the consequence of every assumption made. If such a discussion is thought to be important, or indeed if more detailed information or analysis is required by the reader, appropriate references may be made to the books listed in the Bibliography.

In general, to solve a particular problem, there are a number of numerical techniques available. It is up to a numerical analyst to decide which of these techniques is the most suitable for the particular problem. Assuming that a numerical solution to the problem has been found, a numerical analyst must also be able to decide how meaningful, that is, how accurate this solution is. Although no claim is made that the reader who has worked through this book will be in a position to make these decisions in every case, he should at the very least be keenly aware of these problems. Most techniques discussed in this book are accompanied with some form, even if a rather simplified one, of error analysis. If two or more techniques are suggested for solving a particular problem, either a comparison of the techniques is made or at least a reference to such a discussion is given.

Numerous examples are given to illustrate and explain how the various techniques work. The examples are chosen to be very simple so that the numerical work should not obscure the concepts being described.

Although the book is partly based on a series of lectures given to Mathematics students at Bedford College, University of London, it does not follow any particular syllabus, but nevertheless covers important sections of most undergraduate and National Diploma courses in Numerical Analysis and will also be found useful to sixth form students.

Certain results used in a number of proofs in this book have been proved in the companion volume *Calculus via Numerical Analysis*, reference [3] in the Bibliography.

Although the importance of Numerical Analysis has been recognized by scientists for a long time, it is the widespread use of automatic computers which has made it an indispensable tool in many scientific and most mathematical establishments.

With the rapid progress of technology and the capability of manufacturing ever more sophisticated and reliable components, mathematical modelling is becoming progressively more ambitious. The aim of such models is to express the behaviour of physical systems in terms of mathematical equations. This serves a double purpose: to analyse extremely complex natural phenomena so as to predict their behaviour, and also to synthesize systems whose behaviour is specified. The equations, which may be algebraic, differential or integral must be solved as accurately as possible. It would be very exceptional to be able to solve them by analytical methods; in most cases numerical techniques must be used. The fact that, with the aid of automatic computers, numerical techniques have provided accurate solutions to these equations has in turn led to the formulation of more realistic models and a greater demand for better and more accurate components. It is hardly an exaggeration to claim that in the last twenty-five years we have witnessed the birth of a technological revolution and that Numerical Analysis with the aid of computers is one of its components.

Finally, as far as mathematical education is concerned, it has already been suggested (reference [3]) that Numerical Analysis plays as important a role in mathematics as Calculus does. Indeed these two branches of mathematics are complementary and there are many good reasons why they should be taught side by side at schools and universities.

NOTATION USED

$\mathfrak{R}$	The field of real numbers.
$f(x_i)$ or f_i	A function f evaluated at x_i.
Δ_h or Δ	The forward difference operator.
∇_h or ∇	The backward difference operator.
δ_h or δ	The central difference operator.
E_h or E	The displacement operator.
μ_h or μ	The averaging operator.
D	The differentiation operator.
$\alpha\,(1)\,\beta$	Increase the number α by steps of 1 until the number β is reached.
$\binom{p}{n}$	$\dfrac{p(p-1)(p-2)\ldots(p-n+1)}{n!}$
B_i	Bessel coefficients.
E_i	Everett coefficients.

I FINITE DIFFERENCES

1.1 The forward difference operator

In this chapter we shall be considering functions, such as the function f defined in Table 1, having domain and co-domain $\mathfrak{R}$, the set of real numbers, or a subset of $\mathfrak{R}$. Frequently the domain of the function considered is a discrete set of equally spaced numbers, such as the values of x in Table 1.

TABLE 1

x	f	Δf	$\Delta^2 f$
5·2	140·608		
		8·269	
5·3	148·877		0·318
		8·587	
5·4	157·464		0·324
		8·911	
5·5	166·375		0·330
		9·241	
5·6	175·616		0·336
		9·577	
5·7	185·193		0·342
		9·919	
5·8	195·112		

The set of values $\{x_i\}$ are called *tabular points* and the corresponding values of f are known as *tabular values*. The constant *step-length* (or *interval*) between two consecutive tabular points is generally denoted by h, so that

$$x_{i+1} - x_i = h \quad (i = \ldots, -2, -1, 0, 1, 2, \ldots)$$

In Table 1, $h = 0\cdot1$. The values in the third column of

Table 1, under the heading Δf, are called the *first differences* of f. Any particular first difference is obtained by *subtracting* the appropriate tabular value from the tabular value immediately *below* it, the resulting first difference is then placed on a level midway between these two tabular values. The values in the last column of Table 1 under the heading $\Delta^2 f$ are known as the *second differences* of f. They have the same relation to the first differences as the first differences have to the tabular values. We can extend Table 1 and define the *third* and indeed *higher order differences* of f in the same way as we have defined the first and second order differences.

To generalize the ideas just discussed and to develop a theory of finite differences we need to introduce an appropriate notation.

We generally consider a function f (sometimes we refer to it as the function $y = f(x)$) tabulated at the tabular points

$$\ldots, x_{-2}, x_{-1}, x_0, x_1, x_2, \ldots,$$

the suffix 0 referring to some conveniently labelled origin. The corresponding images or tabular values are labelled

$$\ldots, f_{-2}, f_{-1}, f_0, f_1, f_2, \ldots,$$

where f_i is a shorthand for $f(x_i)$, and, since h is fixed, we write

$$f(x_i + h) = f_{i+1}.$$

When using this notation it is assumed that the tabulated points

$$\ldots, x_{-2}, x_{-1}, x_0, x_1, x_2, \ldots,$$

are arranged in increasing order of magnitude.

In constructing the differences we are defining a new mapping

$$x \mapsto f(x + h) - f(x) \quad (x \in \mathbb{R}),$$

h being the constant interval between two consecutive tabular points. This mapping maps a function f to another function called the *differenced function* of f, and is called an *operator*. Since, as we shall see, h can frequently be chosen to suit the particular problem, we denote this operator by Δ_h. The suffix h is dispensed with when it is quite clear from the context what value of h is chosen—as in the case of Table 1, for example. With this notation we write the first differences as

$$\Delta_h f(x_i) = \Delta_h f_i = f(x_i + h) - f(x_i) \qquad (1.1)$$
$$= f_{i+1} - f_i \quad (i = \ldots, -2, -1, 0, 1, 2, \ldots).$$

The second differences are then written as

$$\Delta_h^2 f_i = \Delta_h(\Delta_h f_i) = \Delta_h f_{i+1} - \Delta_h f_i$$

(where the index 2 does not denote power 2, but indicates the second difference), and, in general, the mth differences (say) as

$$\Delta_h^m f_i = \Delta_h^{m-1}(\Delta_h f_i) = \Delta_h^{m-1} f_{i+1} - \Delta_h^{m-1} f_i. \qquad (1.2)$$

It is instructive to rewrite Table 1 using the above notation. We shall choose the origin to be the tabular point $x_0 = 5 \cdot 5$. Since in the case of Table 1 the step length h is obviously $0 \cdot 1$, we omit the suffix in Δ_h.

TABLE 2

x	f	Δf	$\Delta^2 f$
x_{-3}	f_{-3}		
		Δf_{-3}	
x_{-2}	f_{-2}		$\Delta^2 f_{-3}$
		Δf_{-2}	
x_{-1}	f_{-1}		$\Delta^2 f_{-2}$
		Δf_{-1}	
x_0	f_0		$\Delta^2 f_{-1}$
		Δf_0	
x_1	f_1		$\Delta^2 f_0$
		Δf_1	
x_2	f_2		$\Delta^2 f_1$
		Δf_2	
x_3	f_3		

From the table it is seen that the suffix i $(-3 \leqslant i \leqslant 1)$ in $\Delta^2 f_i$, Δf_i and f_i remains constant along a diagonal line sloping upwards (from right to left). This is the reason why Δ_h is known as the *forward difference* operator.

1.2 Some other operators and the relations between them

We define the operators by the following equations:

The *backward difference operator* ∇_h by

$$\nabla_h f(x) = f(x) - f(x - h). \qquad (1.3)$$

The *central difference operator* δ_h by

$$\delta_h f(x) = f\left(x + \frac{h}{2}\right) - f\left(x - \frac{h}{2}\right). \qquad (1.4)$$

The *displacement operator* E_h by

$$E_h f(x) = f(x + h). \qquad (1.5)$$

The *averaging operator* μ_h by

$$\mu_h f(x) = \tfrac{1}{2}\left[f\left(x + \frac{h}{2}\right) + f\left(x - \frac{h}{2}\right) \right]. \qquad (1.6)$$

The *differentiation operator* D by

$$Df(x) = \frac{df(x)}{dx}. \qquad (1.7)$$

It is fairly simple to find the relationships between the various operators given above. These relations are useful when certain formulae, such as the interpolation formulae, are being established. When using the operator notation we generally omit the suffix h when it is clear from the context what it should be.

The relation between Δ and E

$$
\begin{aligned}
\Delta f(x) &= f(x + h) - f(x) \quad &\text{(by definition of } \Delta) \\
&= Ef(x) - f(x) \quad &\text{(by definition of } E) \\
&= (E - I)f(x), \quad & (1.8)
\end{aligned}
$$

I being the *unit* (or *identity*) *operator*, leaving a function f unchanged. We now write symbolically

$$\Delta = E - I.$$

Although the theory of operators is fascinating in its own right, we shall not discuss the theory in this book. On the other hand, it is important to understand that, for example, in the above formula, we do not subtract the I operator from the E operator (in the usual sense of the word 'subtract'), nor do we interpret the equality sign exactly as we do for algebraic equations, say for $y = 2x$. As pointed out the operator equation is symbolic, and it symbolizes the relationship existing between functions in the codomain (or the image space) of the operators. In other words, the equation tells us that the function Δf is the same as the function $(E - I)f$, so long as we interpret $(E - I)f$ to be

$$(E - I)f(x) = f(x + h) - f(x) \quad (x \in \text{domain of } f).$$

(It is often customary to use the digit 1 (one) instead of the unit operator I, in which case we would write the relation between Δ and E as $\Delta = E - 1$.)

The relation between ∇ and E

To find the relation between ∇ and E and indeed between the various other operators and E, we need to discuss the meaning of $E^p f_i$. By definition

$$Ef_i = f(x_i + h) = f_{i+1},$$

so that

$$E^2 f_i = E(Ef_i) = E(f_{i+1}) = f_{i+2},$$

and so generally we define

$$E^p f_i = f_{i+p}.$$

We assume the above interpretation of E^p for all real p. Since

$$\nabla f(x) = f(x) - f(x - h)$$
$$= f(x) - E^{-1}f(x)$$
$$= (1 - E^{-1})f(x),$$

we obtain

$$\nabla = 1 - E^{-1}. \tag{1.10}$$

The relation between E *and* δ

$$\delta f(x) = f\left(x + \frac{h}{2}\right) - f\left(x - \frac{h}{2}\right)$$
$$= E^{\frac{1}{2}}f(x) - E^{-\frac{1}{2}}f(x)$$
$$= (E^{\frac{1}{2}} - E^{-\frac{1}{2}})f(x).$$

Hence

$$\delta = E^{\frac{1}{2}} - E^{-\frac{1}{2}}. \tag{1.11}$$

The relation between E *and* D

All the operators, except the D operator, are dependent on h. The D operator is independent of h, indeed it is obtained as a result of taking the limit of Δ_h/h as h tends to zero (see reference [3]). Any relation between D and any of the other operators is a relation between finite and infinitesimal mathematics and hence is of great importance. The link is established by the use of the Taylor series. Since in practice it is easier to work with hD rather than with the operator D alone, we shall find the relation between E and hD.

$$Ef(x) = f(x + h)$$
$$= f(x) + hf'(x) + \frac{h^2}{2!} f''(x) + \ldots$$
$$= f(x) + hDf(x) + \frac{h^2}{2!} D^2f(x) + \ldots$$
$$= \left[1 + hD + \frac{h^2 D^2}{2!} + \ldots\right] f(x)$$
$$= e^{hD}f(x).$$

So that

$$E = e^{hD}$$

and hence

$$hD = \ln E. \tag{1.12}$$

We can similarly find relations between the other operators. These are summarized in Table 3 opposite.

We shall develop in this chapter a number of interpolation formulae in terms of the three operators Δ, ∇ and δ. To use

TABLE 3

	E	Δ	∇	δ	hD
E	E	$1+\Delta$	$(1-\nabla)^{-1}$	$1+\dfrac{\delta^2}{2}+\delta\left(1+\dfrac{\delta^2}{4}\right)^{\frac{1}{2}}$	e^{hD}
Δ	$E-1$	Δ	$\nabla(1-\nabla)^{-1}$	$\dfrac{\delta^2}{2}+\delta\left(1+\dfrac{\delta^2}{4}\right)^{\frac{1}{2}}$	$e^{hD}-1$
∇	$1-E^{-1}$	$1-(1+\Delta)^{-1}$	∇	$\delta\left(1+\dfrac{\delta^2}{4}\right)^{\frac{1}{2}}-\dfrac{\delta^2}{2}$	$1-e^{-hD}$
δ	$E^{\frac{1}{2}}-E^{-\frac{1}{2}}$	$\Delta(1+\Delta)^{-\frac{1}{2}}$	$\nabla(1-\nabla)^{-\frac{1}{2}}$	δ	$2\sinh\dfrac{hD}{2}$
hD	$\ln E$	$\ln(1+\Delta)$	$-\ln(1-\nabla)$	$2\sinh^{-1}\delta/2$	hD
μ	$\tfrac{1}{2}(E^{\frac{1}{2}}+E^{-\frac{1}{2}})$	$\left(1+\dfrac{\Delta}{2}\right)(1+\Delta)^{-\frac{1}{2}}$	$\left(1-\dfrac{\nabla}{2}\right)(1-\nabla)^{-\frac{1}{2}}$	$\left(1+\dfrac{\delta^2}{4}\right)^{\frac{1}{2}}$	$\cosh\dfrac{hD}{2}$

these formulae it is necessary to be able to identify the appropriate differences in a difference table. To this end it is advisable to be familiar with the layout of a difference table in terms of the three operators, as in Tables 4, 5 and 6 below.

Forward differences layout

TABLE 4

x	f	Δf	$\Delta^2 f$	$\Delta^3 f$	$\Delta^4 f$	$\Delta^5 f$	$\Delta^6 f$
x_{-3}	f_{-3}						
		Δf_{-3}					
x_{-2}	f_{-2}		$\Delta^2 f_{-3}$				
		Δf_{-2}		$\Delta^3 f_{-3}$			
x_{-1}	f_{-1}		$\Delta^2 f_{-2}$		$\Delta^4 f_{-3}$		
		Δf_{-1}		$\Delta^3 f_{-2}$		$\Delta^5 f_{-3}$	
x_0	f_0		$\Delta^2 f_{-1}$		$\Delta^4 f_{-2}$		$\Delta^6 f_{-3}$
		Δf_0		$\Delta^3 f_{-1}$		$\Delta^5 f_{-2}$	
x_1	f_1		$\Delta^2 f_0$		$\Delta^4 f_{-1}$		
		Δf_1		$\Delta^3 f_0$			
x_2	f_2		$\Delta^2 f_1$				
		Δf_2					
x_3	f_3						

Backward differences layout

TABLE 5

x	f	∇f	$\nabla^2 f$	$\nabla^3 f$	$\nabla^4 f$	$\nabla^5 f$	$\nabla^6 f$
x_{-3}	f_{-3}						
		∇f_{-2}					
x_{-2}	f_{-2}		$\nabla^2 f_{-1}$				
		∇f_{-1}		$\nabla^3 f_0$			
x_{-1}	f_{-1}		$\nabla^2 f_0$		$\nabla^4 f_1$		
		∇f_0		$\nabla^3 f_1$		$\nabla^5 f_2$	
x_0	f_0		$\nabla^2 f_1$		$\nabla^4 f_2$		$\nabla^6 f_3$
		∇f_1		$\nabla^3 f_2$		$\nabla^5 f_3$	
x_1	f_1		$\nabla^2 f_2$		$\nabla^4 f_3$		
		∇f_2		$\nabla^3 f_3$			
x_2	f_2		$\nabla^2 f_3$				
		∇f_3					
x_3	f_3						

$$\text{TABLE } 6$$

x_{-3}	f_{-3}						
		$\delta f_{-5/2}$					
x_{-2}	f_{-2}		$\delta^2 f_{-2}$				
		$\delta f_{-3/2}$		$\delta^3 f_{-3/2}$			
x_{-1}	f_{-1}		$\delta^2 f_{-1}$		$\delta^4 f_{-1}$		
		$\delta f_{-1/2}$		$\delta^3 f_{-1/2}$		$\delta^5 f_{1/2}$	
x_0	f_0		$\delta^2 f_0$		$\delta^4 f_0$		$\delta^6 f_0$
		$\delta f_{1/2}$		$\delta^3 f_{1/2}$		$\delta^5 f_{1/2}$	
x_1	f_1		$\delta^2 f_1$		$\delta^4 f_1$		
		$\delta f_{3/2}$		$\delta^3 f_{3/2}$			
x_2	f_2		$\delta^2 f_2$				
		$\delta f_{5/2}$					
x_3	f_3						

From Table 5 it is seen that the subscripts in ..., $\nabla^3 f$, $\nabla^2 f$, ∇f, f are constant along a diagonal line sloping downwards (from right to left), whereas Table 6 shows that the subscripts are constant along a horizontal line. Note that the odd central differences, those in δf, $\delta^3 f$, $\delta^5 f$, ..., involve fractional subscripts.

Looking at the three tables we note that, for example,

$$\Delta^2 f_{-2} = \nabla^2 f_0 = \delta^2 f_{-1}.$$

Each of these differences is of course the same element in the difference table. We can show this from the definitions of these differences, thus

$$\begin{aligned}
\Delta^2 f_{-2} &= \Delta f_{-1} - \Delta f_{-2} \\
&= (f_0 - f_{-1}) - (f_{-1} - f_{-2}) \\
&= f_0 - 2f_{-1} + f_{-2} \\
\nabla^2 f_0 &= \nabla f_0 - \nabla f_{-1} \\
&= (f_0 - f_{-1}) - (f_{-1} - f_{-2}) \\
&= f_0 - 2f_{-1} + f_{-2}
\end{aligned}$$

and

$$\delta^2 f_{-1} = \delta f_{-1/2} - \delta f_{-3/2}$$
$$= (f_0 - f_{-1}) - (f_{-1} - f_{-2})$$
$$= f_0 - 2f_{-1} + f_{-2}.$$

1.3 Differences of polynomial functions

Up till now we have been differencing functions defined by tables, that is, functions whose domains are discrete sets of values in $\mathbb{R}$. But our definition of the difference operator is much more general and applies to any function whose domain is $\mathbb{R}$ or a subset of $\mathbb{R}$.

Example 1

Find the differenced function of

(a) $f(x) = 3x + 2$ $\qquad$ $(x \in \mathbb{R})$
(b) $g(x) = 2x^2 + x - 1$ $\quad$ $(x \in \mathbb{R})$

for the spacing $h = 1$.

Note that in the example h had to be specified, since for a different h we would obtain different solutions.

Solution

(a) $\Delta_1 f(x) = f(x + 1) - f(x)$
$$= 3(x + 1) + 2 - (3x + 2),$$

so that
$$\Delta_1 f(x) = 3 \quad (x \in \mathbb{R}).$$

(b) $\Delta_1 g(x) = [2(x + 1)^2 + (x + 1) - 1] - [2x^2 + x - 1]$
$$= 4x + 3,$$

so that
$$\Delta_1 g(x) = 4x + 3 \quad (x \in \mathbb{R}).$$

The functions in example 1 are *polynomial functions* of degree 1 and 2 respectively. Any expression of the form

$$p(x) = a_n x^n + a_{n-1} x^{n-1} + \ldots + a_1 x + a_0 \quad (x \in \mathbb{R})$$

20

is called a polynomial function. If $a_n \neq 0$, the function is of *degree n*.

On differencing a polynomial function g of degree 2, we found a polynomial function of degree 1. On differencing a polynomial function f of degree 1 we found a polynomial function of degree 0 (i.e. a *constant function*). It would thus appear that differencing a polynomial function of degree n should result in a polynomial function of degree $n - 1$. This is proved in the following way:

$$\Delta_h p(x) = a_n(x + h)^n + a_{n-1}(x + h)^{n-1} + \ldots +$$
$$+ a_1(x + h) + a_0 -$$
$$- [a_n x^n + a_{n-1} x^{n-1} + \ldots + a_1 x + a_0]$$
$$= a_n[(x + h)^n - x^n] + -a_{n-1}[(x + h)^{n-1} -$$
$$- x^{n-1}] + \ldots + a_1[(x + h) - x].$$

On expansion of the terms in brackets we find that

$$a_n[(x + h)^n - x^n] = \text{polynomial function of degree } n - 1$$

$$a_{n-1}[(x + h)^{n-1} - x^{n-1}]$$
$$= \text{polynomial function of degree } n - 2, \text{ etc.}$$

On adding these results, we obtain:

$$\Delta_h\{\text{polynomial function of degree } n\}$$
$$= \text{polynomial function of degree } n - 1.$$

This result is of fundamental importance in interpolation theory. It implies that on differencing a polynomial function of degree n we obtain a polynomial function of degree $(n - 1)$ which, on differencing, results in a polynomial function of degree $(n - 2)$, and so on until, on differencing n times, we obtain a polynomial function of degree 0. For example, we have seen that if

$$g(x) = 2x^2 + x - 1$$

$$\Delta_1 g(x) = 4x + 3 \qquad (x \in \mathbb{R}),$$

so that $\qquad \Delta_1^2 g(x) = 4 \qquad\qquad (x \in \mathbb{R}).$

Further, $\qquad \Delta_1^3 g(x) = 0 \qquad\qquad (x \in \mathbb{R}).$

We could of course verify this (at least for a reduced domain for x) by forming a difference table. For example, we can evaluate $g(x)$ for $-3 \leqslant x \leqslant 3$, $h = 1$, and form the differences.

TABLE 7

x	$g(x)$	$\Delta_1 g(x)$	$\Delta_1^2 g(x)$	$\Delta_1^3 g(x)$
-3	14			
		-9		
-2	5		4	
		-5		0
-1	0		4	
		-1		0
0	-1		4	
		3		0
1	2		4	
		7		0
2	9		4	
		11		
3	20			

Evaluating $\Delta_1 g(x) = 4x + 3$ for $x = -3 \leqslant x \leqslant 3$, we obtain the values in the third column of Table 7. The second differences are constant and equal to 4, and the third (and higher) differences are all zero as predicted.

1.4 Errors in difference tables

In numerical work we encounter a number of different types of errors, but they can be crudely classified into two categories: *blunders* and *inherent errors*.

Blunders are usually due to human errors. During transcription of data, a frequent blunder is the switching of two digits, for example writing 6398 instead of 6389. Quite frequently if such a blunder occurs in copying values from a book of mathematical tables, or in one reading from a series in a scientific experiment, it can be detected and corrected by the method discussed below.

Inherent errors are due to a number of causes—rounding-off numbers, truncating errors (that is, approximating to the true value of x by the first few terms of an infinite series), and the limitations of measuring instruments in an experiment are just some of them. Very little can be done about eliminating inherent errors, and it is up to the numerical analyst to

TABLE 8

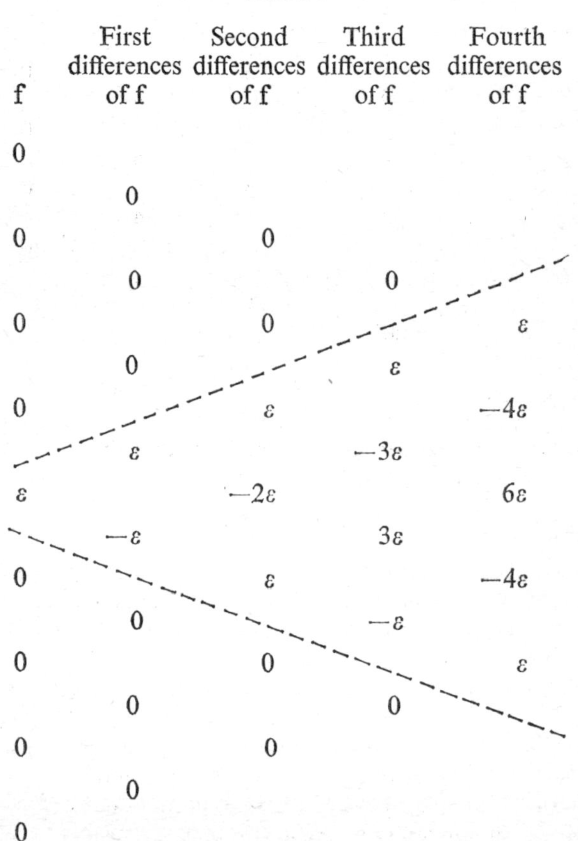

f	First differences of f	Second differences of f	Third differences of f	Fourth differences of f
0				
	0			
0		0		
	0		0	
0		0		
	0		0	
0		0		ε
	0		ε	
0		ε		-4ε
	ε		-3ε	
ε		-2ε		6ε
	$-\varepsilon$		3ε	
0		ε		-4ε
	0		$-\varepsilon$	
0		0		ε
	0		0	
0		0		
	0			
0				

decide whether any particular (computer) calculated result is meaningful or whether it is so distorted by the build-up of inherent errors that it is meaningless. We shall discuss later on in this section the build-up of inherent errors due to round-off in the successive differences in a difference table.

Blunders

We shall consider how to locate and correct a blunder of the type discussed above. We do this by forming a difference table in which we deliberately introduce an error, and then trace the propagation of this error through the table. For simplicity we tabulate the zero function f, taking one of the tabular values to be ε instead of 0.

Table 8 demonstrates the triangular pattern of the affected differences. Note the rapid increase in the magnitude of the maximum error as higher differences are constructed. In the rth differences column the errors are proportional to the coefficients in the binomial expansion of $(1 - x)^r$, and the largest error(s) in a column of differences is along (or on each side of) the horizontal line passing through the error-affected tabular value.

Example 2

Trace and correct the error in a tabular value of the difference table opposite

Convention

It is a convention to drop the decimal points and zeros on the left when calculating the differences. Thus the number 6, the first entry in the $\Delta^3 f$ column in Table 9, means 0·006 whereas −24 means −0·024.

Note how the consecutive entries in each difference column, up to the third differences, decrease in magnitude as the order of the differences increases. In the columns of the third and fourth differences we notice changes in sign which could be explained if there is a blunder in the tabular value corresponding to $x = 1·6$ (compare with Table 8). This

24

TABLE 9

x	f	Δf	$\Delta^2 f$	$\Delta^3 f$	$\Delta^4 f$
1·1	1·331				
		397			
1·2	1·728		72		
		469		6	
1·3	2·197		78		0
		547		6	
1·4	2·744		84		−10
		631		−4	
1·5	3·375		80		40
		711		36	
1·6	4·086		116		−60
		827		−24	
1·7	4·913		92		40
		919		16	
1·8	5·832		108		−10
		1027		6	
1·9	6·859		114		0
		1141		6	
2·0	8·000		120		
		1261			
2·1	9·261				

is confirmed by observation of the fourth differences, where we note that in magnitude the largest value (-60) is on the horizontal line passing through the tabular value corresponding to $x = 1\cdot6$. This particular value is the one most affected by the error (by an amount 6ε as we can see from Table 8). The blunder is thus located, and it must now be corrected. We assume that the differences of order k, where k is determined by inspection, are constant except for the propagated error. From Table 9 it is reasonable to assume that the third differences should be constant, all equal to a, say. Under these circumstances we have (compare with Table 8):

$$0\cdot036 - (-0\cdot024) = (a - 3\varepsilon) - (a + 3\varepsilon),$$

25

so that
$$\varepsilon = -0{\cdot}10.$$

It follows that

the affected entry $= 4{\cdot}086 = $ (correct entry) $+ \varepsilon$,

hence

$$\text{the (correct entry)} = 4{\cdot}086 - \varepsilon$$
$$= 4{\cdot}096.$$

Using this corrected value, we obtain the following difference table:

TABLE 10

x	f	Δf	$\Delta^2 f$	$\Delta^3 f$
1·1	1·331			
		397		
1·2	1·728		72	
		469		6
1·3	2·197		78	
		547		6
1·4	2·744		84	
		631		6
1·5	3·375		90	
		721		6
1·6	4·096		96	
		817		6
1·7	4·913		102	
		919		6
1·8	5·859		108	
		1027		6
1·9	6·859		114	
		1141		6
2·0	8·000		120	
		1261		
2·1	9·261			

Table 10 is of course a difference table for a cubic polynomial function f.

Inherent errors

To examine the effect on successive differences of round-off errors in the tabular values, we shall consider the worst possible case—when the round-off errors are maxima. For a particular tabular value the maximum round-off error is half a unit in the last decimal place. The greatest effect of these errors in the difference table occurs when they alternate in sign in the successive tabular values, as in Table 11.

TABLE 11

Maximum round-off error	First differences	Second differences	Third differences	Fourth differences
$\frac{1}{2}$				
	-1			
$-\frac{1}{2}$		2		
	1		-4	
$\frac{1}{2}$		-2		8
	-1		4	
$-\frac{1}{2}$		2		-8
	1		-4	
$\frac{1}{2}$		-2		
	-1			
$-\frac{1}{2}$				

It is clear from Table 11 that the build-up of errors due to round-off can be very rapid. In the case of table 11 the magnitude of the errors is doubled with each differencing, so that the error in the nth differences is 2^{n-1}. It follows that when working with (say) four decimal places, the maximum error in the rth differences due to round-off is

$$2^{r-1} \times 10^{-4}.$$

For example, Table 12 is a difference table for the polynomial function of degree 3

$$f : x \mapsto 2x^3 + 1,$$

the tabular values being evaluated correct to four decimal places.

TABLE 12

x	$f(x)$	$\Delta f(x)$	$\Delta^2 f(x)$	$\Delta^3 f(x)$
2·00	17·0000			
		0·2412		
2·01	17·2412		24	
		0·2436		0
2·02	17·4848		24	
		0·2460		2
2·03	17·7308		26	
		0·2486		−4
2·04	17·9794		22	
		0·2508		
2·05	18·2302			

In this case

$$\Delta^3 f : x \mapsto 12h^3,$$

and since $h = 0 \cdot 01$, the third differences should be constant and equal to 12×10^{-6}. The magnitude of the maximum possible error due to round-off is

$$4 \times 10^{-4}$$

and is many times greater than the actual third differences. It follows that the numbers in the column of the third differences in Table 12 are due to the propagation of the round-off errors and not the third differences of the function f. This explains the erratic changes in both the magnitudes and the signs of these numbers. It is extremely important in problems involving function approximation and in interpolation to decide whether the differences of some order under investigation are constant except for the inherent error propagation. The above discussion is the basis on which a decision can be made.

Finally, how does one distinguish between blunders and inherent errors? Basically the answer lies in the magnitude of the errors. Since the *maximum* round-off error is half a unit in the last decimal place, whereas the *minimum* blunder is one unit in the last decimal place, there should be no real difficulty in general in distinguishing between the two types of error.

1.5 Interpolation formulae

Given a number of tabulated points say $x_1, x_2, \ldots, x_6$ and the corresponding tabular values $f_1, f_2, \ldots, f_6$, we wish to find the polynomial function of the lowest degree passing through these points:

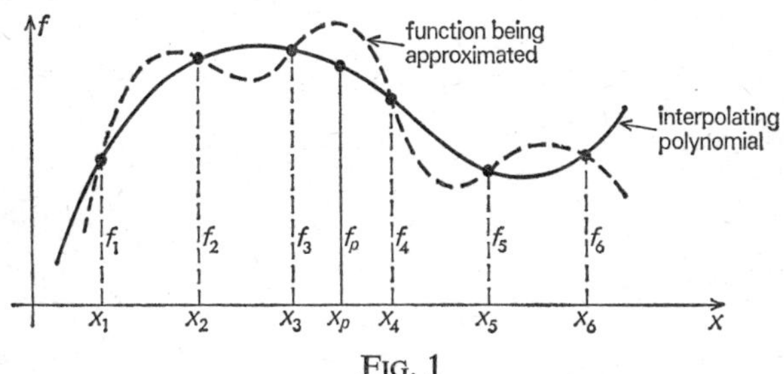

FIG. 1

Having found such a polynomial function, we can evaluate it for any x in the interval of $\mathfrak{R}$ defined by the end tabular points. We call such a polynomial an *interpolation formula* and by *interpolation* we mean the evaluation of the polynomial for some such value of x.

In the particular case when the tabular points are separated by a constant step length, finding an interpolation formula is fairly straightforward.

The Gregory–Newton forward difference interpolation formula

Let $x_p = x_0 + ph$, where h is the constant step length and p is a real number. The value x_0 is a conveniently chosen tabular point. We have

$$f_p = f(x_0 + ph) = E^p f_0.$$

29

Using the relation between E and Δ (see Table 3, page 17), we obtain:

$$f_p = (1 + \Delta)^p f_0.$$

In expanding this, we use the notation

$$\binom{p}{n} = \frac{p(p-1)(p-2)\ldots(p-n+1)}{n!}\,.$$

Note that $\binom{p}{n}$ is a polynomial of degree n in p.

We now write

$$f_p = (1 + \Delta)^p f_0 = f_0 + \binom{p}{1}\Delta f_0 + \binom{p}{2}\Delta^2 f_0 +$$

$$+ \binom{p}{3}\Delta^3 f_0 + \ldots + \binom{p}{n}\Delta^n f_0 + R_{n+1}(p), \quad (1.13)$$

where

$$R_{n+1}(p) = \binom{p}{n+1}\Delta^{n+1}f_0 + \binom{p}{n+2}\Delta^{n+2}f_0 + \ldots$$

If $R_{n+1}(p) = 0$, the formula for f_p is known as *the interpolating polynomial of degree n*. It is the unique polynomial of degree n passing through the $n+1$ tabular points $x_0, x_1, \ldots, x_n$ at which it takes the values $f_0, f_1, \ldots, f_n$ (see section 2.1).

If we sample some function at $n + 1$ points, we can always find the interpolating polynomial having the same values as the sampled function at these points, but, of course there is no guarantee that they will have any other points in common or indeed that the interpolating polynomial will be a good approximation to the function in between the sampled (tabular) points (see Fig. 1). Indeed it is not at all clear under what circumstances the formula we have found for f_p is valid. Fortunately, so long as the function we wish to approximate is continuous, Weierstrass's approximation theorem (see reference [6]) guarantees that it can be approximated by some polynomial to any desired accuracy over any specified interval. Although this theorem does not help in finding the interpolating polynomial, it does give us some confidence in using the interpolating formula. Equation (1.13) also gives us the difference $|R_{n+1}(p)|$ between the function being approximated and the interpolating polynomial. For any particular value of p we can generally obtain an idea of what this difference is (we generally call it the *error*) by calculating the first term of the series for $|R_{n+1}(p)|$.

Finally, *interpolation* means calculating the (approximate) value of a function at a particular point x of its domain. It follows that we do not need to find the actual interpolating polynomial but to evaluate it at the particular point of interest.

Example 3

From the difference Table, Table 13, calculate (a) f(1·2) and (b) f(2·5):

TABLE 13

x	$f(x)$	$\Delta f(x)$	$\Delta^2 f(x)$
0	1		
		-1	
1	0		4
		3	
2	3		4
		7	
3	10		4
		11	
4	21		4
		15	
5	36		4
		19	
6	55		

Solution

(a) We choose $x_0 = 1$.

$$x_p = 1\cdot2 = x_0 + ph, \text{ where } h = 1.$$

It follows that $p = 0.2$, so that

$$f(1.2) = f_0 + 0.2\Delta f_0 + \frac{0.2(0.2 - 1)}{2}\Delta^2 f_0$$

$$= 0 + 0.2 \times 3 + \frac{0.2 \times (-0.8)}{2} \times 4.$$

$$= 0.28$$

(b) Choose $x_0 = 2$.

$$xp = 2.5 = x_0 + ph, \text{ so that } p = 0.5.$$

$$f(2.5) = 3 + 0.5 \times 7 + \frac{0.5(0.5 - 1)}{2} \times 4$$

$$= 6.$$

Note that since the second differences are constant, the function being approximated is itself a polynomial of degree 2. In fact, it is the function

$$f: x \mapsto 2x^2 - 3x + 1 \quad (x \in \mathbb{R}).$$

In this case, the interpolated values are of course exact.

Had we attempted to find $f(5.5)$ from Table 13, we could not have used the forward difference interpolation formula. In this case we would need to use the backward difference formula which we now discuss.

The Gregory–Newton backward difference interpolation formula

We establish this formula in a similar way to the forward difference formula except that this time we express the operator E in terms of the backward difference operator ∇ instead of the forward difference operator Δ.

$$f_p = f(x_0 + ph) = E^p f_0 = (1 - \nabla)^{-p} f_0$$

$$= f_0 + p\nabla f_0 + \frac{p(p + 1)}{2!}\nabla^2 f_0 + \ldots +$$

$$+ \frac{p(p + 1)\ldots(p + n - 1)}{n!}\nabla^n f_0 +$$

$$+ R_{n+1}(p), \quad (1.14)$$

where

$$R_{n+1}(p) = \frac{p(p + 1)\ldots(p + n)}{(n + 1)!}\nabla^{n+1} f_0 + \ldots.$$

If we assume that $R_{n+1}(p) = 0$, the equation for f_p is a polynomial of degree n. Since there is only one polynomial of degree n passing through the $(n + 1)$ points $x_0, x_1, \ldots, x_n$, at which it has values $f_0, f_1, \ldots, f_n$, it follows that the backward difference and the forward difference interpolating polynomials are one and the same. The distinction is that written in one form, the forward difference form, the polynomial is useful for interpolation at the top of a table, whereas in the backward difference form it is useful for interpolation at the bottom of the table.

Example 4

From the difference table, Table 13, find $f(5 \cdot 5)$.

Solution

A reference to Table 5 (page 18) for a reminder of the layout of backward differences will be useful at this stage.

We choose in Table 13, $x_0 = 5$.

$$x_p = x_0 + ph, \text{ so that } p = 0 \cdot 5.$$

$$f(5 \cdot 5) = f_0 + p \nabla f_0 + \frac{p(p + 1)}{2!} \nabla^2 f_0 + \ldots$$

$$= 36 + 0 \cdot 5 \times 15 + 0 \cdot 375 \times 4$$

$$= 45.$$

Examples 3 and 4 demonstrate the usefulness of the forward difference formula for interpolation at a point near the top of a difference table and of the backward difference formula for interpolation at a point near the bottom of the difference table. But as a general rule it is a good policy to arrange the difference table, and choose the point x_0 in such a way that the interpolation should be at some point near the middle of the difference table. To understand this we need to consider more closely the error term in interpolation.

We have derived the interpolation formula equation (1.14) in the form

$$f = Q_n + R_{n+1},$$

where Q_n is the interpolating polynomial of degree n and $|R_{n+1}|$ is the error. Since by construction, f and Q_n are equal at the tabular points $x_0, x_1, \ldots, x_n$, it follows that R_{n+1} must be zero at these points. Thus R_{n+1} must have the following form:

$$R_{n+1} = (x - x_0)(x - x_1) \ldots (x - x_n)S(x) \quad (1.16)$$
$$= q_{n+1}(x)S(x) \quad \text{(say)},$$

where $S(x)$ is some function of x and

$$q_{n+1} = (x - x_0) \ldots (x - x_n).$$

Since $x = x_p = x_0 + ph$, and $x_{i+1} - x_i = h$ (for all i), it follows that

$$q_{n+1} = p(p - 1) \ldots (p - n)h^{n+1}.$$

Below are graphs of some of the q_{n+1} functions.

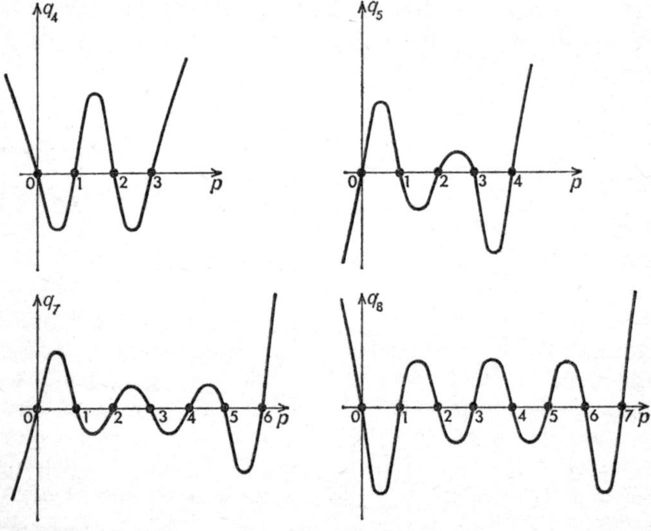

FIG. 2

It is clear from the diagrams that in general $|q_{n+1}|$ is smaller when x is chosen in the middle rather than at an end of an

34

interval. This means that for least error we should interpolate at a point near the middle of the interval considered,
and for this purpose we use one of the *central difference
formulae*. Two of the most widely used central difference
formulae are the *Everett* and the *Newton–Bessel* formulae.
To find the Newton–Bessel formula, an assumption is made
that f_p can be expressed in the form:

$$f_p = (a_0 + a_1\delta^2 + a_2\delta^4 + \ldots)f_0 + \\ + (b_0 + b_1\delta^2 + b_2\delta^4 + \ldots)f_1,$$

where $a_0, a_1, \ldots b_0, b_1, \ldots$ are constants. The details of how
these constants are determined can be found in most books
on Numerical Analysis, for example in reference [1]. The
formula is

$$f_p = f_0 + B_1\delta f_{\frac{1}{2}} + B_2(\delta^2 f_0 + \delta^2 f_1) + B_3\delta^3 f_{\frac{1}{2}} + \\ + B_4(\delta^4 f_0 + \delta^4 f_1) + B_5\delta^5 f_{\frac{1}{2}} + \ldots \quad (1.17)$$

where
$$B_1 = p,$$

$$B_2 = \tfrac{1}{2}\frac{p(p-1)}{2!},$$

$$B_3 = \frac{p(p-\frac{1}{2})(p-1)}{3!},$$

$$B_4 = \tfrac{1}{2}\frac{(p+1)p(p-1)(p-2)}{4!},$$

$$B_5 = \frac{(p+1)p(p-\frac{1}{2})(p-1)(p-2)}{5!}.$$

Note that
$$B_i(p) = B_i(1-p) \text{ when } i \text{ is even}$$
and
$$B_i(p) = -B_i(1-p) \text{ when } i \text{ is odd},$$

so that, for example,
$$B_2(0{\cdot}3) = B_2(0{\cdot}7),$$
$$B_3(0{\cdot}3) = -B_3(0{\cdot}7)$$
and
$$B_3(0{\cdot}5) = B_5(0{\cdot}5) = \ldots = 0.$$

The Bessel coefficients B_2, B_3, . . . are cumbersome to evaluate. Fortunately comprehensive sets of tables are available (see reference [2]), where B_2, B_3 and B_4 are tabulated for $p = 0(0 \cdot 001)1$.

The Newton–Bessel formula is particularly suitable for interpolation in the range $0 < p < 1$, and it is probably the most used of all interpolation formulas. Another frequently used central-difference formula is the Everett formula; it is in fact a rearrangement of the Newton-Bessel formula.

The Everett interpolation formula

This formula is established in most books on Numerical Analysis (see reference [1]). It is

$$f_p = f_0 + p\delta f_{\frac{1}{2}} + E_2\delta^2 f_0 + E_4\delta^4 f_0 + \ldots +$$
$$+ F_2\delta^2 f_1 + F_4\delta^4 f_1 + \ldots, \quad (1.18)$$

where

$$E_2 = \frac{-p(p-1)(p-2)}{3!},$$

$$E_4 = -\frac{(p+1)p(p-1)(p-2)(p-3)}{5!},$$

$$F_2 = \frac{(p+1)p(p-1)}{3!},$$

$$F_4 = \frac{(p+2)(p+1)p(p-1)(p-2)}{5!}.$$

E_2, E_4, . . . F_2, F_4, . . . are known as *Everett coefficients* and, as with Bessel coefficients, they are available in tables (see reference [2]).

Note that

$$E_i(p) = F_i(1-p) \text{ for } i = 2, \ 4$$

(this is in fact true for $i = 2, 4, 6, \ldots$), so that, for example,

$$E_2(0 \cdot 2) = F_2(0 \cdot 8) \text{ and } E_2(0 \cdot 8) = F_2(0 \cdot 2).$$

A particular advantage of the Everett formula is that differences of odd order, apart from the first, are not used and therefore do not need to be tabulated.

Example 5

Use the Newton–Bessel formula (equation 1.17) to find f(0·25) from the following table:

TABLE 14

x	$f(x)$	First differ- ences	Second differ- ences	Third differ- ences	Fourth differ- ences
0	0		(—)	(—)	
		79 656			
0·1	0·079 656		793		
		78 863		766	
0·2	0·158 519		1559		41
		77 304		725	
0·3	0·235 823		2284		71
		75 020		654	
0·4	0·310 843		2938		79
		72 082		575	
0·5	0·382 925		3513		
		68 569			
0·6	0·451 494				

Solution

We choose
$$x_0 = 0·2, \quad h = 0·1$$
and
$$x_0 + ph = 0·25,$$
so that
$$p = 0·5$$
and
$$B_1 = 0·5000,$$
$$B_2 = -0·0625,$$
$$B_3 = 0,$$
$$B_4 = 0·0117.$$

It follows that

$$f(0·25) = 0·158\ 519 + 0·5 \times 0·077\ 304 -$$
$$- 0·0625 \times (-0·001\ 559 - 0·002\ 284) + 0 +$$
$$+ 0·0117 \times (0·000\ 041 + 0·000\ 071)$$
$$= 0·197\ 413.$$

Example 6

Use the Everett formula to calculate f(0·312 56) from Table 14.

$$f_p = (1 - p)f_0 + E_2\delta^2 f_0 + E_4\delta^4 f_0 + \ldots +$$
$$+ pf_1 + F_2\delta^2 f_1 + F_4\delta^4 f_1 + \ldots .$$

Note that

$$f_0 + p\delta f_{\frac{1}{2}} = f_0 + p(f_1 - f_0) = (1 - p)f_0 + pf_1.$$

In this case we choose $x = 0·3$ and since $h = 0·1$ and $ph = 0·012\ 56$ we find that $p = 0·1256$. We find then:

$$E_2 = -0·0342, \quad F_2 = -0·0206,$$
$$E_4 = 0·0055, \quad F_4 = 0·0041,$$

so that

$$f_p = 0·8744 \times 0·235\ 823 + (-0·0342) \times (-0·002\ 284) +$$
$$+ 0·0055 \times 0·0071 + \ldots +$$
$$+ 0·1256 \times 0·310\ 843 + (-0·0206) \times$$
$$\times (-0·002\ 938) + 0·0041 \times 0·000\ 007\ 9 + \ldots$$
$$\text{i.e. } f_p = 0·245\ 385.$$

1.6 The throwback technique

Both the central difference formulae discussed above are used in practice to interpolate a large number of tabular points, and frequently in making tables of tabulated values. In these cases it is important to do the calculations in the most economical way possible. The throwback technique helps in this objective for a limited (but nevertheless the most useful) interval of p, from 0 to 1.

Neglecting the fifth and higher order differences, the technique consists of modifying the second differences in such a way that the fourth differences in the formula can also be neglected. This will be illustrated in the case of the Newton–Bessel formula.

For $0 < p < 1$, the ratio

$$\frac{B_4}{B_2} = \frac{(p + 1)(p - 2)}{12}$$

remains nearly constant, in fact in this interval the ratio varies between $-\frac{1}{6}$ and $-\frac{3}{16}$. In these circumstances it seems reasonable to replace $B_4(\delta^4 f_0 + \delta^4 f_1)$ by $CB_2(\delta^4 f_0 + \delta^4 f_1)$ where C is chosen so as to minimise the resulting error. Starting with the truncated Newton–Bessel formula

$$f_p = f_0 + p\delta f_{\frac{1}{2}} + B_2(\delta^2 f_0 + \delta^2 f_1) + B_3 \delta^3 f_{\frac{1}{2}} + B_4(\delta^4 f_0 + \delta^4 f_1),$$

we want to find the value for C for which

$$|f_p - f_p^{(1)}| = \varepsilon$$

is minimized, where

$$f_p^{(1)} = f_0 + p\delta f_{\frac{1}{2}} + B_2[(\delta^2 + C\delta^4)f_0 + \\ + (\delta^2 + C\delta^4)f_1] + B_3 \delta^3 f_{\frac{1}{2}},$$

and is sometimes known as the *Bessel formula with modified differences*. On subtraction, we find

$$\varepsilon = |(B_4 - CB_2)(\delta^4 f_0 + \delta^4 f_1)|.$$

To minimize ε, we must find the value of C for which $q(p) = B_4 - CB_2$ is a minimum in the interval $0 < p < 1$.

$$q(p) = \tfrac{1}{48}[(p + 1)p(p - 1)(p - 2) - \tfrac{1}{4}Cp(p - 1)]$$
$$= \tfrac{1}{48}[p^4 - 2p^3 - p^2 + 2p - 12C(p^2 - p)].$$

The stationary values of q occur when

$$(p - \tfrac{1}{2})[(p^2 - p - 1) - 6C] = 0,$$

i.e. when $p = \frac{1}{2}$ or when $p = \frac{1}{2} \pm \sqrt{\frac{5}{4} + 6C}$. When $p = \frac{1}{2}$, q has a maximum.

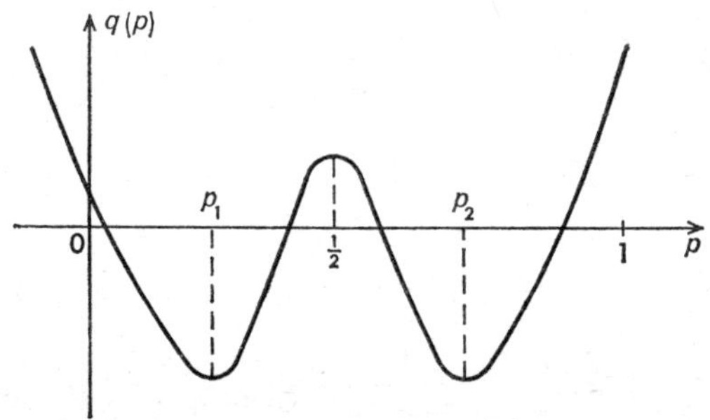

Fig. 3

Let

$$p_1 = \tfrac{1}{2} - \sqrt{\tfrac{5}{4} + 6C} \quad \text{and} \quad p_2 = \tfrac{1}{2} + \sqrt{\tfrac{5}{4} + 6C}.$$

From a sketch of the graph of q, it is clear that this function has minima at $p = p_1$ and $p = p_2$. In fact these minima are of equal magnitude, since at both these values of p,

$$q(p) = \frac{-1}{48}(1 + 6C)^2.$$

Also from the sketch it is clear that we should choose C so that the extreme values of q are minimized. For this purpose we choose

$$q(p_1) = -q(\tfrac{1}{2}).^*$$

But

$$q(\tfrac{1}{2}) = \tfrac{1}{48}(\tfrac{9}{16} + 3C),$$

so that the above condition implies that

$$(1 + 6C)^2 = \tfrac{9}{16} + 3C,$$

hence

$$C = -0{\cdot}066 \quad \text{or} \quad C = -0{\cdot}184.$$

The value $C = -0{\cdot}066$ is inadmissible, since (for example) it would mean that p_2 is outside the range $[0, 1]$.

It follows that the best choice for C is

$$C = -0{\cdot}184.$$

We write

$$(\delta^2 - 0{\cdot}184\delta^4)f_0 \quad \text{as} \quad \delta_m^2 f_0,$$

* This choice of equal extrema for 'best approximation' can in fact be proved rigorously. The proof is based on the following theorem:

The best polynomial approximation of degree n to an arbitrary single valued function f in the interval $[a, b]$ is p such that
$$\varepsilon(x) = f(x) - p(x)$$
attains a greatest maximum value m at not less than $(n + 2)$ distinct points in $[a, b]$, and has alternatively the values m and $-m$ at these points.

The proof of this theorem is given in several textbooks, for example see reference [1], p. 137. A more rigorous and extensive treatment of 'best approximation' is found in reference [6].

and

$$(\delta^2 - 0{\cdot}184\delta^4)f_1 \quad \text{as} \quad \delta_m^2 f_1.$$

These new differences are called *modified second differences*. The Newton–Bessel formula with modified differences becomes

$$f_p = f_0 + p\delta f_{\frac{1}{2}} + B_2(\delta_m^2 f_0 + \delta_m^2 f_1) + B_3\delta^3 f_{\frac{1}{2}}, \quad (1.19)$$

and this device of modifying the differences is known as *throwback*. With this choice of C,

$$|q(p)| < 0{\cdot}23 \times 10^{-3}.$$

It follows that so long as the fourth order differences are less than about 1000, the modified differences formula gives very accurate results in the interval $[0, 1]$.

A similar argument applies to the throwback device for the Everett formula. It will be appreciated that the throwback technique is basically a method of approximating one polynomial function (of degree five in this case) by another polynomial function of lower degree (of degree four in this case). Such approximations can in general be very accurate over certain small regions of the domain.

Example 7

Use the modified Newton–Bessel formula, equation (1.19), to find $f(0{\cdot}25)$ from Table 14 (also see example 5).

Solution

$$f_p = f_0 + p\delta f_{\frac{1}{2}} + B_2(\delta_m^2 f_0 + \delta_m^2 f_1) + B_3\delta^3 f_{\frac{1}{2}}$$

$$\begin{aligned}
\delta_m^2 f_0 + \delta_m^2 f_1 &= (\delta^2 f_0 + \delta^2 f_1) - 0{\cdot}184(\delta^4 f_0 + \delta^4 f_1) \\
&= -0{\cdot}003\ 843 - 0{\cdot}184 \times 0{\cdot}000\ 112 \\
&= -0{\cdot}003\ 864.
\end{aligned}$$

$$\begin{aligned}
f(0{\cdot}25) &= 0{\cdot}148\ 419 + 0{\cdot}038\ 652 + 0{\cdot}000\ 242 \\
&= 0{\cdot}197\ 413.
\end{aligned}$$

1.7 Numerical differentiation and integration

Both differentiation and integration formulae will be

developed from the interpolation formulae which have been established. But although these integration formulae are in general extremely accurate, the differentiation ones can involve very large errors. This is due to the fact that both the differentiation and integration formulae are found by differentiating and integrating the interpolating polynomials. Integration, involving finding the area under the curve, is not very sensitive to a small error in a tabular value. Differentiation on the other hand, being the slope of the curve at the point in question, is extremely sensitive to such an error. Indeed, the derivative at x_0 of a function f is approximately

$$\frac{f_1 - f_0}{h} \quad \text{(see reference [3])}$$

and this approximation is all the better the smaller the tabular spacing $h = x_1 - x_0$. But, assuming that f_1 and f_0 are given subject to some relative error, the smaller the value of h, the greater is the relative error in the difference $(f_1 - f_0)$. Hence h must not be chosen very small. On the other hand if h is not small, the derivative approximation is not very good. In practice the optimal value for the spacing h is usually found by trial and error.

1.8 Some formulae for differentiation

Forward difference formula

From Table 3 we find the relationship:

$$D = \frac{1}{h} \ln E = \frac{1}{h} \ln (1 + \Delta)$$

$$= \frac{1}{h} (\Delta - \tfrac{1}{2}\Delta^2 + \tfrac{1}{3}\Delta^3 - \tfrac{1}{4}\Delta^4 + \ldots).$$

Thus

$$f_0' = Df_0 = \frac{1}{h} (\Delta f_0 - \tfrac{1}{2}\Delta^2 f_0 + \tfrac{1}{3}\Delta^3 f_0 - \tfrac{1}{4}\Delta^3 f_0 + \ldots). \quad (1.20)$$

The equivalent backward difference formula is

$$f_0' = \frac{1}{h} (\nabla f_0 + \tfrac{1}{2}\nabla^2 f_0 + \tfrac{1}{3}\nabla^3 f_0 + \tfrac{1}{4}\nabla^4 f_0 + \ldots). \quad (1.21)$$

These two formulae, equations (1.20) and (1.21), are rarely used since the convergence is relatively slow.

Central difference formulae

To establish differentiation formulae in terms of central differences we can use any of the central differences interpolation forms. By differentiating them repeatedly, we obtain derivatives of various orders.

For example, since

$$f_p = f_0 + p\delta f_{\frac{1}{2}} + B_2(\delta^2 f_0 + \delta^2 f_1) + \\ + B_3\delta^3 f_{\frac{1}{2}} + B^4(\delta^4 f_0 + \delta^4 f_1) + \ldots,$$

and

$$f_p' = \frac{\mathrm{d}f_p}{\mathrm{d}x} = \frac{\mathrm{d}f_p}{\mathrm{d}p}\cdot\frac{\mathrm{d}p}{\mathrm{d}x} = \frac{1}{h}\frac{\mathrm{d}f_p}{\mathrm{d}p} \quad (x = x_0 + ph),$$

it follows that

$$hf_p' = \frac{\mathrm{d}f_p}{\mathrm{d}p} = \delta f_{\frac{1}{2}} + B_2'(\delta^2 f_0 + \delta^2 f_1) + \\ + B_3'\delta^3 f_{\frac{1}{2}} + B_3'(\delta^4 f_0 + \delta^4 f_1) + \ldots, \quad (1.22)$$

and (after simplification)

$$h^2 f_p'' = \frac{\mathrm{d}^2 f_p}{\mathrm{d}p^2} = \delta^2 f_0 + p\delta^3 f_{\frac{1}{2}} + \\ + B_4''(\delta^4 f_0 + \delta^4 f_1) + \ldots, \quad (1.23)$$

and similarly for higher derivatives. B_2', B_3', $\ldots$ are the derivatives of B_2, B_3, $\ldots$ with respect to p. For example, since

$$B_2 = \tfrac{1}{4}p(p - 1),$$
$$B_2' = \tfrac{1}{2}(p - \tfrac{1}{2}),$$

and since

$$B_3 = \tfrac{1}{6}(p^2 - \tfrac{1}{2}p)(p - 1),$$
$$B_3' = \tfrac{1}{12}(6p^2 - 6p + 1)$$

and $\qquad\qquad B_3'' = p - \tfrac{1}{2}.$

The derivatives B_3', B_4', B_5', B_6' and B_3'', B_4'', B_5'', B_6' are available for $p \in [0, 1]$ in tables (see for example reference [2]).

The above formulae for derivatives, equations (1.22) and (1.23),' converge more rapidly than the forward and backward difference formula equations (1.20) and (1.21). They are also quite general since they enable us to calculate the various derivatives at non-tabular points.

Since the errors involved in numerical differentiation are large, rather than use the above formulae it is sometimes better to find the derivatives at the tabular points only, and to interpolate for derivatives at the non-tabular points. There are many formulae for this purpose. For example, we can use again a relation between operators from Table 3:

$$hD = 2 \sinh^{-1} \delta/2.$$

Now, an expansion for $\sinh^{-1} x$ is

$$\sinh^{-1} x = x - \frac{x^3}{6} + \frac{3x^5}{40} - \frac{5x^7}{112} + \ldots, \quad (|x| < 1)$$

so that

$$hf_0' = (2 \sinh^{-1} \delta/2)f_0$$
$$= \delta f_0 - \tfrac{1}{24}\delta^3 f_0 + \ldots.$$

This series for f_0' is seen to be in terms of *odd order* differences δf_0, $\delta^3 f_0$, ... But, as can be seen in Table 6, these differences do not appear in the difference table. To overcome this difficulty we introduce the operator μ in the following way:

$$hf_0' = \frac{1}{\mu} (2 \sinh^{-1} \delta/2)\mu f_0$$

$$= \left(1 + \frac{\delta^2}{4}\right)^{-\frac{1}{2}} (2 \sinh^{-1} \delta/2)\mu f_0$$

$$= (1 - \tfrac{1}{8}\delta^2 + \tfrac{3}{128}\delta^4 - \ldots)$$
$$(\delta - \tfrac{1}{24}\delta^3 + \tfrac{3}{640}\delta^5 - \ldots)\mu f_0$$
$$= (\mu\delta - \tfrac{1}{6}\mu\delta^3 + \tfrac{1}{30}\mu\delta^5 - \ldots)f_0. \tag{1.24}$$

Formulae for second and all even order derivatives are easier to find. For example,

$$h^2 f_0'' = h^2 D^2 f_0 = (2 \sinh^{-1} \delta/2)^2 f_0$$
$$= (\delta^2 - \tfrac{1}{12}\delta^4 + \tfrac{1}{90}\delta^6 - \ldots)f_0. \tag{1.25}$$

Similar techniques are used to establish higher order derivatives.

Example 8

From the difference table below calculate f'(0·77) and f''(0·77).

TABLE 15

x	$f(x)$	δf	$\delta^2 f$	$\delta^3 f$	$\delta^4 f$
0·73	0·796 586				
		25 731			
0·75	0·822 317		329		
		26 060		10	
0·77	0·848 377		339		1
		26 399		11	
0·79	0·874 776		350		
		26 749			
0·81	0·901 525				

Solution

$$hf'(0·77) = \mu\delta f(0·77) - \tfrac{1}{6}\mu\delta^3 f(0·77) + \ldots$$

$$= \tfrac{1}{2}[0·026\ 060 + 0·026\ 399] -$$
$$- \tfrac{1}{12}[0·000\ 010 + 0·000\ 011] + \ldots$$

$$= 0·026\ 229 - 0·000\ 001 = 0·026\ 228.$$

Since $h = 0·02$,

$$f'(0·77) = \frac{0·026\ 228}{0·02} = 1·3114.$$

Similarly,

$$h^2 f''(0·77)\ \delta^2 f(0·77) - \tfrac{1}{12}\delta^4 f(0·077) + \ldots$$

$$= 0·000\ 339 - \tfrac{1}{12} \times 10^{-6}$$

$$= 0·000\ 339.$$

It follows that

$$f''(0·77) = 0·8475.$$

In this example the tabulated function was sinh and the two derivatives are

$$\cosh (0 \cdot 77) = 1 \cdot 311\ 390$$

and

$$\sinh (0 \cdot 77) = 0 \cdot 848\ 377.$$

1.9 Integration

As already mentioned, numerical integration unlike differentiation is inherently an accurate operation. This can be explained intuitively in the following way. We wish to find the integral $\int_{x_1}^{x_3} f$, that is, to calculate the area under the curve drawn in a continuous line in Figure 4 between x_1 and x_3.

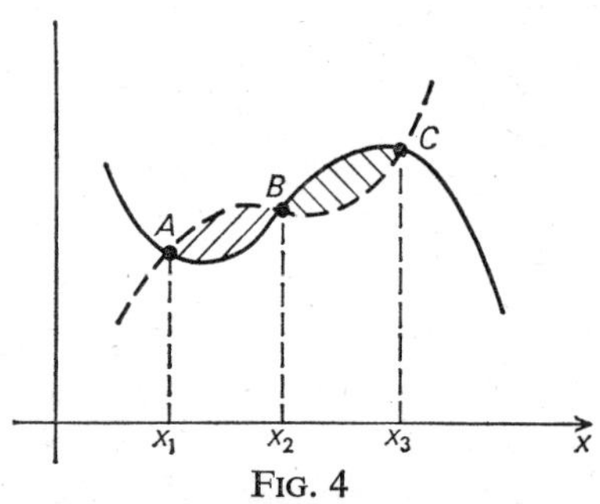

Fig. 4

The interpolated curve passing through the three tabular points is shown by the discontinuous line in the figure. In numerical integration we find the area under the interpolated curve. Although between x_1 and x_2 this implies overestimating the area shown by the shaded region between A and B, the calculated area from x_2 to x_3 is underestimated by an amount shown by the shaded region between B and C. The two shaded regions are approximately equal, so that the calculated area is a very accurate approximation to the actual area. It follows that in similar circumstances, numerical integration can be a particularly accurate operation. Whenever possible problems which require differentiation should

46

be re-examined and if possible set out in a form which requires integration instead of differentiation.

Simpson's rule

We begin with the first three terms of the forward difference interpolation formula, i.e.

$$f = f_0 + p\Delta f_0 + \tfrac{1}{2}(p^2 - p)\Delta^2 f_0,$$

and integrate over the interval x_0 to x_2. Observing that, when

$$x = x_0 + ph$$

then

$$\int f \, dx = h \int f \, dp,$$

it follows that

$$\int_{x_0}^{x_2} f \, dx = h \int_0^2 (f_0 + p\Delta f_0 + \tfrac{1}{2}(p^2 - p)\Delta^2 f_0) \, dp$$

$$= h[2f_0 + 2\Delta f_0 + \tfrac{1}{3}\Delta^2 f_0],$$

i.e.
$$\int_{x_0}^{x_2} f \, dx = \frac{h}{3}[f_0 + 4f_1 + f_2], \tag{1.26}$$

since $\Delta f_0 = f_1 - f_0$ and $\Delta^2 f_0 = f_2 - 2f_1 + f_0$.

Equation (1.26) is known as *Simpson's rule*. At first sight it may appear that having truncated the interpolation formula after only three terms, we cannot except good accuracy from the resulting integration formula. This is not so, since, if we consider the fourth term of the interpolation formula $\binom{p}{3} \Delta^3 f_0$, its contribution to the integral is

$$h \int_0^2 \tfrac{1}{6}(p^3 - 3p^2 + 2p)\Delta^3 f_0 \, dp = 0.$$

Since the $\Delta^3 f_0$ term does not contribute to the integral, Simpson's rule is an efficient method of integration.

The efficiency of Simpson's rule is demonstrated above since, although it was established on the basis of interpolating with a polynomial of degree two, it is shown that the same rule holds good for an interpolating polynomial of degree three. The accuracy can be increased by taking h sufficiently small, and this is done in the following way. Let us assume that we wish to integrate f in the interval $[a, b]$. We divide this interval into an *even* number n of sub-intervals so that

$$h = \frac{b - a}{n}.$$

We then have

$$\int_{x_0 = a}^{x_n = b} f \, dx = \int_{x_0}^{x_2} + \int_{x_2}^{x_4} + \ldots + \int_{x_{n-2}}^{x_n} f \, dx$$

$$= \frac{h}{3} (f_0 + 4f_1 + f_2) + \frac{h}{3} (f_2 + 4f_3 + f_4) +$$

$$+ \ldots + \frac{h}{3} (f_{n-2} + 4f_{n-1} + f_n).$$

Thus,

$$\int_{x_0}^{x_n} f \, dx = \frac{n}{3} [f_0 + 2(f_2 + f_4 + \ldots + f_{n-2}) +$$

$$+ 4(f_1 + f_3 + \ldots + f_{n-1}) + f_n].$$

In this form Simpson's rule can be used to evaluate the integral in any specified interval $[a, b]$ and allows the choice of h to be compatible with the requirements of accuracy of the problem being considered.

Since Simpson's rule is such a widely used technique of integration it is worth while to discuss its accuracy at this stage. The inherent error of the method is due to truncating the interpolation formula after the fourth term. Now by the repeated application of the first mean value theorem of calculus (see reference [3]), we have

$$\Delta f_0 = f_1 - f_0 = hf'(v_1), \quad \text{where } x_0 \leqslant v_1 \leqslant x_1;$$

$$\Delta^2 f_0 = h^2 f''(v_2), \quad \text{where } x_0 \leqslant v_2 \leqslant x_2;$$

$$\Delta^3 f_0 = h^3 f'''(v_3), \quad \text{where } x_0 \leqslant v_3 \leqslant x_3;$$

$$\Delta^4 f_0 = h^4 f^{iv}(v_4), \quad \text{where } x_0 \leqslant v_4 \leqslant x_4;$$

etc.

Since h is small it follows that the dominant term in the truncated remainder of the interpolation formula is $\binom{p}{4}\Delta^4 f_0$, and its contribution to the integral is the approximate error involved in this numerical process.

The contribution is

$$h \int_0^2 \binom{p}{4} \Delta^4 f_0 \, dp = -\frac{h}{90} \Delta^4 f_0 = -\frac{h^5}{90} f^{iv}(v).$$

In fact it can be shown that v can be found such that

$$h^5 f^{iv}(v) \int_0^2 \binom{p}{4} dp$$

is equal to the total remainder of the truncated interpolation formula. (The interested reader is referred to reference [3] where Simpson's rule and the associated errors are established with the aid of Taylor series. See also the discussion in Section 2.4.) It follows that

$$-\frac{h^5}{90} f^{iv}(v)$$

is the error associated with Simpson's rule. It is seen that for small h, the error is extremely small, so that this technique of integration is very accurate. The simplicity and accuracy of the method has resulted in it being very widely used in numerical calculations.

Example 9

Use Simpson's integration formula with $h = 0.1$ to evaluate

$$\int_0^{0.6} \frac{dx}{1 + x^2},$$

and check the result from tables. Work with numbers rounded to five decimal places.

Solution

We first tabulate the function

$$f(x) = \frac{1}{1 + x^2} \quad \text{for} \quad x = 0{\cdot}1(0{\cdot}1)\,0{\cdot}6.$$

x	x_0 0	x_1 0·1	x_2 0·2	x_3 0·3	x_4 0·4	x_5 0·5	x_6 0·6
$f(x)$	1·0000	0·990 10	0·961 54	0·917 43	0·862 07	0·800 00	0·735 29

Now apply Simpson's rule,

$$\int_0^{0\cdot6} f\,\mathrm{d}x = \frac{h}{3}\left[f_0 + 2(f_2 + f_4) + 4(f_1 + f_3 + f_5) + f_6\right]$$

$$= \frac{0\cdot1}{3}\left[1\cdot000\ 00 + 2 \times (1\cdot823\ 61) + \right.$$
$$\left. + 4 \times (2\cdot707\ 53) + 0\cdot735\ 29\right]$$

$$= \frac{0\cdot1}{3} \times 16\cdot212\ 63$$

$$= 0\cdot540\ 42 \ \text{(to five decimal places)}.$$

$$Check: \quad \int_0^{0\cdot6} \frac{\mathrm{d}x}{1 + x^2} = \left[\arctan x\right]_0^{0\cdot6} = 0\cdot540\ 420.$$

Some other integration formulae
The forward difference formula

Integrating the forward difference interpolation formula in the interval $[x_0, x_1]$, we obtain

$$\int_{x_0}^{x_1} f\,\mathrm{d}x = h \int_0^1 \left[f_0 + \binom{p}{1}\Delta f_0 + \binom{p}{2}\Delta^2 f_0 + \ldots\right]\mathrm{d}p$$

$$= h[f_0 + \tfrac{1}{2}\Delta f_0 - \tfrac{1}{12}\Delta^2 f_0 + \tfrac{1}{24}\Delta^3 f_0 - \tfrac{19}{720}\Delta^4 f_0 + \ldots].$$

The backward difference formula

$$\int_{x_0}^{x_1} f\,\mathrm{d}x = f \int_0^1 \left[f_0 + p\nabla f_0 + \tfrac{1}{2}(p^2 + p)\nabla^2 f_0 + \right.$$

$$+ \tfrac{1}{6}(p^3 + 3p^2 + 2p)\nabla^3 f_0 + \ldots]\,\mathrm{d}p$$

$$= h[f_0 + \tfrac{1}{2}\nabla f_0 + \tfrac{5}{12}\nabla^2 f_0 + \tfrac{3}{8}\nabla^3 f_0 + $$

$$+ \tfrac{251}{720}\nabla^4 f_0 + \ldots]. \quad (1.27)$$

These two formulae are rarely used for evaluating integrals,

50

but they are sometimes used in the solution of differential equations (Section 4.4). The last equation, in particular, is a well known *predictor* formula and is used in conjunction with a *corrector* obtained by integrating the backward difference formula in the interval $[x_0, x_1]$ where x_0 is chosen to correspond to $p = -1$, so that x_1 corresponds to $p = 0$. We then obtain

$$\int_{x_0}^{x_1} f \, dx = h \int_{-1}^{0} (f_0 + p\nabla f_0 - \tfrac{1}{2}(p^2 + p)\nabla^2 f_0 + \ldots) \, dp$$

$$= h[f_0 - \tfrac{1}{2}\nabla f_0 - \tfrac{1}{12}\nabla^2 f_0 \ldots].$$

Since the subscript of f is arbitrary, we can advance it by 1 to obtain

$$\int_{x_0}^{x_1} f \, dx = h[f_1 - \tfrac{1}{2}\nabla f_1 - \tfrac{1}{12}\nabla^2 f_1 - \ldots]. \quad (1.28)$$

(The next two terms in the brackets are

$$-\tfrac{1}{24}\nabla^3 f_1 - \nabla\tfrac{19}{720}{}^4 f_1.)$$

This is the form in which this formula is used as a corrector.

Bessel's integration formula

Integrating Bessel's interpolation formula over the interval $[x_0, x_1]$, we obtain

$$\int_{x_0}^{x_1} f \, dx = h \int_{0}^{1} [f_0 + p\delta f_{\frac{1}{2}} + \tfrac{1}{2}p(p - 1)\delta^2 f_{\frac{1}{2}} + \ldots] \, dp$$

$$= h(\mu f_{\frac{1}{2}} - \tfrac{1}{12}\mu\delta^2 f_{\frac{1}{2}} + \ldots).$$

(The next two terms in brackets are

$$\tfrac{11}{720}\mu\delta^4 f_0 - \tfrac{191}{60480}\mu\delta^6 f_{\frac{1}{2}}.)$$

The accuracy of this formula depends of course on the number of terms remaining after truncating the series on the right-hand side. For example, if we truncate all terms involving the third and higher order differences, the dominant error term will be of order $O(h^4)$. Again the accuracy of integration will depend on the value of h chosen. If we wish to integrate over a number of intervals, from x_0 to x_n say, we proceed as follows.

$$\int_{x_0}^{x_1} f \, dx = h(\mu f_{\frac{1}{2}} - \tfrac{1}{12}\mu\delta^2 f_{\frac{1}{2}} + \ldots)$$

$$\int_{x_1}^{x_2} f \, dx = h(\mu f_{\frac{3}{2}} - \tfrac{1}{12}\mu\delta^2 f_{\frac{3}{2}} + \ldots)$$

$$\vdots$$

$$\int_{x_{n-1}}^{x_n} f \, dx = h(\mu f_{n-\frac{1}{2}} - \tfrac{1}{12}\mu\delta^2 f_{n-\frac{1}{2}} + \ldots).$$

On adding, we obtain

$$
\begin{aligned}
\int_{x_0}^{x_n} f \, dx &= h(\mu f_{\frac{1}{2}} + \mu f_{\frac{3}{2}} + \ldots + \mu f_{n-\frac{1}{2}}) - \\
&\quad - \tfrac{1}{12}(\mu\delta^2 f_{\frac{1}{2}} + \mu\delta^2 f_{\frac{3}{2}} + \ldots + \mu\delta^2 f_{n-\frac{1}{2}}) + \ldots \\
&= h\{[(\tfrac{1}{2}f_1 + \tfrac{1}{2}f_0) + (\tfrac{1}{2}f_2 + \tfrac{1}{2}f_1) + \ldots + \\
&\quad + (\tfrac{1}{2}f_n + \tfrac{1}{2}f_{n-1})] - \tfrac{1}{12}[\mu(\delta f_1 - \delta f_0) + \\
&\quad + \mu(\delta f_2 - \delta f_1) + \ldots + \mu(\delta f_n - \delta f_{n-1})] + \ldots\}.
\end{aligned}
$$

Hence

$$
\begin{aligned}
\int_{x_0}^{x_n} f \, dx = h[&(\tfrac{1}{2}f_0 + f_1 + \ldots + f_{n-1} + \tfrac{1}{2}f_n) - \\
&- \tfrac{1}{12}(\mu\delta f_n - \mu\delta f_0) + \ldots]. \quad (1.29)
\end{aligned}
$$

(The next term in brackets is

$$\tfrac{11}{720}(\mu\delta^3 f_n - \mu\delta^3 f_0))$$

If the differences of first and higher orders are neglected, this formula reduces to the well known *trapezoidal rule*. The practical difficulty in using integration formulae derived from interpolation formulae is that in general to integrate in the interval $[x_0, x_n]$ we must know the tabular values of the function outside this interval. For example, the last formula uses $\mu\delta f_n$, $\mu\delta^3 f_n$, . . .; these must be found from a difference table. To construct this difference table we need the tabular values f_{n+1}, f_{n+2}, f_{n+3}, . . . outside the range of integration. If these tabular values are unknown, we cannot use the previous integration formulae. For this purpose we must develop a formula which uses information within the difference table only—this is the attraction of *Gregory's formula*. Before discussing this formula we need to define a number of *inverse operators*.

Since $Df = f'$, the reverse operation of D, denoted by $\frac{1}{D}$ or D^{-1}, is defined by

$$D^{-1}f'(x) = f(x).$$

From the Fundamental Theorem of Calculus (see reference [3]) we know that D^{-1} is the *indefinite integral operator*. Since $D^{-1}f$ is an indefinite integral with an associated constant of integration, it follows that D and D^{-1} are not commutative, that is, that

$$DD^{-1} \neq D^{-1}D.$$

For example, if $f(x) = 2x + 1$,

$DD^{-1}f(x) = 2x + 1$, but

$D^{-1}Df(x) = 2x + $ arbitrary constant of integration.

Since $\Delta_h f(x) = f(x + h) - f(x) = \psi(x)$ (say), we define Δ^{-1} by the equation

$$\Delta^{-1}\psi(x) = f(x).$$

Note that Δ^{-1} is not a one-one mapping, since if $p(x)$ is some periodic function of period h we also have

$$\Delta[f(x) + p(x)] = \psi(x),$$

so that

$$\Delta^{-1}\psi(x) = f(x) + p(x).$$

This last relation also means that the two operators Δ and Δ^{-1} are not commutative.

The operator Δ^{-1} is sometimes called the *indefinite summation operator*. ∇^{-1} is defined in a similar manner. We

obtain the relation between Δ^{-1} and ∇^{-1} from Table 3. We have
$$\Delta = E\nabla,$$
it follows that
$$\nabla^{-1} = E\Delta^{-1}$$
so that
$$\nabla^{-1}f_n = \Delta^{-1}f_{n+1}.$$

Finally, we shall make use of the expression $\Delta D^{-1}f$. We can interpret it in the following way. Let

$$D^{-1}f = \int f\,dx = F(x) + C \quad \text{say, } C \text{ being a constant.}$$

Then

$$\begin{aligned}
\Delta D^{-1}f &= \Delta(F(x) + C)\\
&= F(x + h) - F(x)\\
&= \int_{x_i}^{x_{i+1}} f\,dx\\
&= \frac{1}{D}f_{i+1} - \frac{1}{D}f_i \quad (i = \ldots, -2, -1, 0, 1, \ldots).
\end{aligned}$$

Gregory's formula

By the results of the previous paragraph we can express the definite integral over the interval $[x_0, x_n]$ as

$$\int_{x_0}^{x_n} f\,dx = \frac{1}{D}f_n - \frac{1}{D}f_0.$$

From the relations between the operators in Table 3, we have
$$hD = \ln(1 + \Delta) \quad \text{and} \quad hD = -\ln(1 - \nabla),$$
so that
$$hD = \Delta - \tfrac{1}{2}\Delta^2 - \tfrac{1}{3}\Delta^3 - \tfrac{1}{4}\Delta^4 + \ldots,$$
and
$$hD = \nabla + \tfrac{1}{2}\nabla^2 + \tfrac{1}{3}\nabla^3 + \tfrac{1}{4}\nabla^4 + \ldots.$$
By simple manipulations, we find that
$$\frac{1}{hD} = \Delta^{-1} + \tfrac{1}{2} - \tfrac{1}{12}\Delta + \tfrac{1}{24}\Delta^2 - \ldots$$

and
$$\frac{1}{hD} = \nabla^{-1} - \tfrac{1}{2} - \tfrac{1}{12}\nabla - \tfrac{1}{24}\nabla^2 - \ldots \ .$$

It now follows that we can write the definite integral as:

$$\int_{x_0}^{x_n} f \, dx = \frac{h}{hD} f_n - \frac{h}{hD} f_0$$

$$= h(\nabla^{-1} - \tfrac{1}{2} - \tfrac{1}{12}\nabla - \tfrac{1}{24}\nabla^2 - \ldots)f_n -$$
$$- h(\Delta^{-1} + \tfrac{1}{2} - \tfrac{1}{12}\Delta + \tfrac{1}{24}\Delta^2 - \ldots)f_0$$

$$= h[(\nabla^{-1}f_n - \Delta^{-1}f_0) - \tfrac{1}{2}(f_0 + f_n) -$$
$$- (\tfrac{1}{12}\nabla f_n - \Delta f_0) - \tfrac{1}{24}(\nabla^2 f_n + \Delta^2 f_0) - \ldots].$$

Since

$$\Delta f_0 = f_1 - f_0$$
$$\Delta f_1 = f_2 - f_1$$
$$\vdots$$
$$\Delta f_n = f_{n+1} - f_n,$$

we have

$$\Delta(f_0 + f_1 + \ldots + f_n) = f_{n+1} - f_0,$$

so that

$$f_0 + f_1 + \ldots + f_n = \Delta^{-1}f_{n+1} - \Delta^{-1}f_0$$
$$= \nabla^{-1}f_n - \Delta^{-1}f_0.$$

Thus,

$$\int_{x_0}^{x_n} f \, dx = h\{(\tfrac{1}{2}f_0 + f_1 + \ldots + f_{n-1} + \tfrac{1}{2}f_n) -$$
$$- \tfrac{1}{12}(\nabla f_n - \Delta f_0) - \tfrac{1}{24}(\nabla^2 f_n + \Delta^2 f_0) - \ldots\}. \quad (1.30)$$

In this form equation (1.30) is known as *Gregory's formula*.

Example 10

Use Gregory's formula to find $\int_{0.70}^{0.73} f \, dx$, given the following information:

$$\begin{array}{cc} x & f(x) \\ 0{\cdot}70 & 0{\cdot}496\ 585 \\ 0{\cdot}71 & 0{\cdot}491\ 644 \\ 0{\cdot}72 & 0{\cdot}486\ 752 \\ 0{\cdot}73 & 0{\cdot}481\ 909 \end{array}$$

Solution

We first form the difference table:

TABLE 16

x		First differences $(-)$	Second differences
0·70	0·496 585		
		4941	
0·71	0·491 644		49
		4892	
0·72	0·486 752		49
		4743	
0·73	0·481 909		

Substituting the numerical data into Gregory's formula, we obtain

$$\int_{0{\cdot}70}^{0{\cdot}73} f\,dx = \tfrac{1}{100}\{(0{\cdot}248\ 293 + 0{\cdot}491\ 644 + 0{\cdot}486\ 752 +$$
$$+\ 0{\cdot}240\ 955) - \tfrac{1}{12}(-0{\cdot}004\ 743 +$$
$$+\ 0{\cdot}004\ 941) - \tfrac{1}{24}(0{\cdot}000\ 049 +$$
$$+\ 0{\cdot}000\ 049)\}$$

$$= \tfrac{1}{100}(1{\cdot}467\ 644 - 0{\cdot}000\ 016 + 0{\cdot}000\ 004)$$
$$= 0{\cdot}014\ 676 \quad \text{(to six decimal places)}.$$

In this example, the function f considered was e^{-x}.

$$\int_{0{\cdot}70}^{0{\cdot}73} e^{-x}\,dx = -[e^{-x}]_{0{\cdot}70}^{0{\cdot}73} = 0{\cdot}496\ 585 - 0{\cdot}481\ 909$$
$$= 0{\cdot}014\ 676.$$

II LAGRANGE INTERPOLATION POLYNOMIAL AND DIVIDED DIFFERENCES

In Chapter I we discussed a number of techniques for interpolating when the tabular points are equidistant. In this chapter we shall consider the more general problem of interpolation when the tabular points x_i $(i = 0, 1, 2, \ldots, n)$ are not necessarily equidistant. As in Chapter I, the problem is to estimate the image of the function at some point x, given its images $f_0, f_1, \ldots, f_n$ at distinct points $x_0, x_1, \ldots, x_n$.

2.1 Lagrange interpolation polynomial

We wish to find the polynomial f passing through the $(n + 1)$ non-equidistant and distinct tabular points $x_0, x_1, \ldots, x_n$ at which the corresponding images are $f_0, f_1, \ldots, f_n$. We assume that this polynomial has the form

$$f(x) = a_0 + a_1 x + a_2 x^2 + \ldots + a_n x^n. \qquad (2.1)$$

We first show that a unique polynomial of degree n passes through $(n + 1)$ distinct points. Indeed let us assume that there are two such polynomials, say h and g, and that they are not equal. Then the polynomial h − g is at most of degree n and is zero for $(n + 1)$ values of x, i.e. for $x = x_i$ $(i = 0, 1, \ldots, n)$. But there is no (non-zero) polynomial satisfying these conditions, hence h = g $(x \in \mathbb{R})$ and so the polynomial is unique.

The Lagrange polynomial $L(x)$ is the (unique) polynomial of degree n whose images at $x_0, x_1, \ldots, x_n$ are $f_0, f_1, \ldots, f_n$. We can find this polynomial in a number of ways, possibly the least complicated method is described below.

We define a polynomial $p(x)$ of degree $(n+1)$ as the product

$$p(x) = \prod_{i=0}^{n} (x - x_i),$$

and we also define (for $k = 0, 1, \ldots, n$)

$$p_k(x) = (x - x_0) \ldots (x - x_{k-1})(x - x_{k+1}) \ldots (x - x_n)$$

$$= \prod_{\substack{i=0 \\ i \neq k}}^{n} (x - x_i).$$

Note that the polynomials $p_k(x)$ ($k = 0, 1 \ldots, n$) are all of degree n. The obvious relation between $p(x)$ and $p_k(x)$ is

$$p(x) = (x - x_k)p_k(x) \quad (k = 0, 1, \ldots, n).$$

Since

$$p_k(x_r) = 0 \qquad \text{when } r \neq k$$
$$= p_k(x_k) \neq 0 \quad \text{when } r = k,$$

it follows that the polynomial of degree n

$$\frac{p_k(x)}{p_k(x_k)} f_k = 0 \qquad \text{when } x = x_r \neq x_k$$

$$= f_k \quad \text{when } x = x_k.$$

When we sum the $(n+1)$ polynomials

$$\frac{p_0(x)}{p_0(x_0)} f_0, \frac{p_1(x)}{p_1(x_1)} f_1, \ldots, \frac{p_n(x)}{p_n(x_n)} f_n,$$

we obtain a polynomial of degree n whose images at x_0, $x_1, \ldots, x_n$ are $f_0, f_1, \ldots, f_n$. By the uniqueness property this must be the polynomial $L(x)$ which we are seeking, i.e.

$$L(x) = \sum_{k=0}^{n} \frac{p_k(x)}{p_k(x_k)} f_k = \sum_{k=0}^{n} L_k(x)f_k, \qquad (2.2)$$

where

$$L_k(x) = \frac{p_k(x)}{p_k(x_k)}$$

$$= \frac{(x - x_0) \ldots (x - x_{k-1})(x - x_{k+1}) \ldots (x - x_n)}{(x_k - x_0) \ldots (x_k - x_{k-1})(x_k - x_{k+1}) \ldots (x_k - x_n)}$$

$$(2.3)$$

$L(x)$ is known as the *Lagrangian interpolation polynomial of degree n* and $L_k(x)$ are the *Lagrangian coefficient functions.* If we use $L(x)$ to approximate a function $f(x)$ in some interval $[a, b]$, then in this interval

$$f(x) = L(x) + R, \qquad\qquad (2.4)$$

where R is the *remainder term* indicating how good the approximation is. In Section 2.2 it will be shown that

$$R = \frac{f^{(n+1)}(\gamma)}{(n+1)!}\, p(x), \text{ where } \gamma \in [a, b]. \qquad (2.5)$$

Example 11

Find the Lagrangian interpolation polynomial relevant to the following data:

x	0	1	3
$f(x)$	8	7	23

Solution

We first compute the Lagrangian coefficients using equation (2.3).

$$L_0(x) = \frac{(x-1)(x-3)}{(0-1)(0-3)} = \tfrac{1}{3}(x^2 - 4x + 3),$$

$$L_1(x) = \frac{(x-0)(x-3)}{(1-0)(1-3)} = -\tfrac{1}{2}(x^2 - 3x),$$

$$L_2(x) = \frac{(x-0)(x-1)}{(3-0)(3-1)} = \tfrac{1}{6}(x^2 - x).$$

Now using equation (2.2), we obtain

$$L(x) = \tfrac{8}{3}(x^2 - 4x + 3) - \tfrac{7}{2}(x^2 - 3x) + \tfrac{23}{6}(x^2 - x)$$
$$= 3x^2 - 4x + 8.$$

Of course, as in the case of finite differences, when interpolating for some specific value of x, we do not need to find the actual polynomial, but only its value at x. Thus in example 11 to interpolate at $x = \frac{1}{2}$ we need to find

$$L_0(\tfrac{1}{2}) = \tfrac{5}{12} \quad L_1(\tfrac{1}{2}) = \tfrac{5}{8} \quad L_2(\tfrac{1}{2}) = -\tfrac{1}{24},$$

so that

$$L(\tfrac{1}{2}) = \tfrac{5}{12} \times 8 + \tfrac{5}{8} \times 7 - \tfrac{1}{24} \times 23$$
$$= 6{\cdot}75.$$

As already mentioned, the Lagrangian interpolation method is more general than the finite difference methods since it is applicable to data given at non-equally spaced intervals. But even for equally spaced intervals it has the advantage of not requiring a difference table.

On the other hand it has a number of disadvantages. For example when deciding upon the degree of the polynomial to use for interpolation, it is fairly simple to estimate the truncation error involved from a difference table (see for example equations (1.13) and (1.14)). This is not so simple in the case of the Lagrangian method unless of course $f(x)$ is given analytically in which case equation (2.5) gives the necessary information.

The analysis of the effects of inherent errors due to the rounding of data is fairly simple in the case of Lagrangian interpolation. This problem arises when we are interpolating for values of a function between entries in a mathematical table for that function, these are of course rounded to a specified number of decimal places. Even if the function can be evaluated exactly at the tabular points, the storage of these values in a computer implies rounding off. The maximum rounding error in the tabular value $f(x_r)$ (say) is half a unit in the last decimal point. So if $f(x_r)$ is correctly rounded to r decimal places, the maximum error in interpolation for $f(x)$ is:

$$\tfrac{1}{2} \times 10^{-r-1} \sum_{k=0}^{n} |L_k(x)|. \qquad (2.6)$$

The reader interested in further exploration of this quantity is advised to refer to a more advanced textbook on this subject (for example, see reference [4]).

2.2 Divided differences

Let $x_0, x_1, \ldots, x_n$ be $(n+1)$ tabular points (not necessarily equidistant) and $f(x_0), f(x_1), \ldots, f(x_n)$ the corresponding tabular values. We define the *first divided difference* of f between x_i and x_{i+1}, denoted by $f(x_i, x_{i+1})$, as

$$f(x_i, x_{i+1}) = \frac{f(x_{i+1}) - f(x_i)}{x_{i+1} - x_i} \quad (i = 0, 1, \ldots, n). \quad (2.7)$$

Note that as a result of this definition

$$f(x_i, x_{i+1}) = f(x_{i+1}, x_i) \quad (i = 0, 1, \ldots n). \quad (2.8)$$

Similarly the *second divided difference* of f between x_i and x_{i+2} is defined as:

$$f(x_i, x_{i+1}, x_{i+2}) = \frac{f(x_{i+1}, x_{i+2}) - f(x_i, x_{i+1})}{x_{i+2} - x_i} \quad (2.9)$$

$$(i = 0, 1, \ldots, n-1).$$

and the *nth divided difference* is

$$f(x_0, x_1, \ldots, x_n)$$
$$= \frac{f(x_1, x_2, \ldots, x_n) - f(x_0, x_1, \ldots, x_{n-1})}{x_n - x_0}. \quad (2.10)$$

From these definitions we can construct a *divided difference table* in a similar manner to that used in constructing finite difference tables. The divided difference table for $(x_0, f_0), \ldots, (x_3, f_3)$ will look as follows:

TABLE 17

x_0	f_0			
		$f(x_0, x_1)$		
x_1	f_1		$f(x_0, x_1, x_2)$	
		$f(x_1, x_2)$		$f(x_0, x_1, x_2, x_3)$
x_2	f_2		$f(x_1, x_2, x_3)$	
		$f(x_2, x_3)$		
x_3	f_3			

The tabular points and the corresponding tabular values are

written in the first two columns. The first divided differences are written in the third column, the second divided differences are written in the fourth column etc.

Example 12

Form the divided difference table for the following data:

$$(-3, 13), (-1, -3), (0, -5), (2, 3).$$

Solution

TABLE 18

-3	13			
		-8		
-1	-3		2	
		-2		0
0	-5		2	
		4		
2	3			

(1) We have

$$f(x_0, x_1) = \frac{f(x_1) - f(x_0)}{x_1 - x_0} = \frac{f(x_0)}{x_0 - x_1} + \frac{f(x_1)}{x_1 - x_0},$$

and

$$f(x_0, x_1, x_2) = \frac{f(x_1, x_2) - f(x_0, x_1)}{x_2 - x_0}$$

$$= \frac{f(x_0)}{(x_0 - x_1)(x_0 - x_2)} + \frac{f(x_1)}{(x_1 - x_0)(x_1 - x_2)} +$$

$$+ \frac{f(x_2)}{(x_2 - x_0)(x_2 - x_1)},$$

and in general (as can be proved by induction):

$$f(x_0, x_1, \ldots, x_k) = \frac{f(x_0)}{(x_0 - x_1) \ldots (x_0 - x_k)} +$$

$$+ \ldots + \frac{f(x_k)}{(x_k - x_0) \ldots (x_k - x_{i-1})}. \tag{2.11}$$

(2) If the tabular points are equidistant, so that

$$x_{i+1} - x_i = h \quad (i = 0, 1, \ldots, n - 1),$$

then

$$f(x_0, x_1) = \frac{\Delta f(x_0)}{h} = \frac{\nabla f(x_1)}{h} = \frac{\delta f(x_{1/2})}{h},$$

and in general

$$f(x_0, x_1, \ldots, x_n) = \frac{\Delta^n f(x_0)}{n!\, h^n}$$
$$= \frac{\nabla^n f(x_n)}{n!\, h^n} = \frac{\delta^n f(x_{n/2})}{n!\, h^n}. \qquad (2.12)$$

(3) By the first mean value theorem (see reference [3])

$$f(x_0, x_1) = \frac{f(x_1) - f(x_0)}{x_1 - x_0} = f'(\mu),$$

where μ is some point in the interval $[x_0, x_1]$, and in general

$$f(x_0, x_1, \ldots, x_n) = \frac{f^{(n)}(\mu)}{n!} \quad (\mu \in [x_0, x_n]) \qquad (2.13)$$

as is shown in Section 2.3 below.

2.3 Newton's interpolation formula with divided differences

Let x be any point in the interval $[x_0, x_n]$, and $f(x)$ the corresponding image under f.

From the definitions of the divided differences, we have

$$f(x, x_0) = \frac{f(x) - f(x_0)}{x - x_0},$$

so that

$$f(x) = f(x_0) + (x - x_0)f(x, x_0),$$

and

$$f(x, x_0, x_1) = \frac{f(x, x_0) - f(x_0, x_1)}{x - x_1},$$

so that

$$f(x, x_0) = f(x_0, x_1) + (x - x_1)f(x, x_0, x_1),$$

and in general,

$$f(x, x_0, \ldots, x_{n-1}) = f(x_0, x_1, \ldots, x_n) +$$
$$+ (x - x_n)f(x, x_0, \ldots, x_n).$$

From these relations we obtain:

$$f(x) = f(x_0) + (x - x_0)f(x_0, x_1) +$$
$$+ (x - x_0)(x - x_1)f(x_0, x_1, x_2) + \ldots +$$
$$+ (x - x_0)(x - x_1)\ldots(x - x_{n-1})f(x_0, x_1, \ldots, x_n) +$$
$$+ R(x), \quad (2.14)$$

where

$$R(x) = (x - x_0)(x - x_1)\ldots(x - x_n)f(x, x_0, x_1, \ldots, x_n).$$

This formula is written as

$$f(x) = p(x) + R(x).$$

Since for given tabular points $x_0, x_1, \ldots, x_n$, factors of the form $f(x, x_1, \ldots, x_r)$ are constants (see Table 18) it follows that $p(x)$ is the sum of polynomials of degrees $0, 1, \ldots, n$ respectively, hence it is a polynomial of degree n, known as the *Newton's interpolation formula with divided differences.* Since $R(x)$ which is known as the *error term*, is zero at $x = x_0$ and x_1 and $\ldots x_n$, the polynomial $p(x)$ agrees with $f(x)$ at these $(n + 1)$ points. But there is a unique polynomial of degree n having this property, so it follows that $p(x)$ must be the Lagrange polynomial discussed in Section 2.1 in terms of divided differences. This time we also have an explicit error term $R(x)$ which gives at any point x in the domain the difference between $f(x)$ and the polynomial $p(x)$. If the first $(n + 1)$ derivatives of the function f are continuous in the interval of interest, say $[a, b]$ we can write this error term in a more useful form. We have

$$f(x) - p(x) = R(x) = \pi(x)f(x, x_0, x_1, \ldots, x_n),$$

where

$$\pi(x) = (x - x_0)(x - x_1)\ldots(x - x_n). \quad (2.15)$$

Since $R(x) = 0$ at the $(n + 1)$ points $x = x_0, x_1, \ldots, x_n$, $f(x) - p(x) = 0$ at these points. By the repeated application of the first mean value theorem, it follows that

$$f^{(n)}(\mu) = p^{(n)}(\mu) \quad \text{for some } \mu \in [a, b]. \quad (2.16)$$

But as mentioned above, p(x) is the sum of $(n + 1)$ polynomials of degrees $0, 1, 2, \ldots, n$ respectively. Hence the nth derivative of p(x) is

$$n!\, f(x_0, x_1, \ldots, x_n)$$

as can be seen from equation (2.14). From equation (2.16), it follows that

$$f(x_0, x_1, \ldots, x_n) = \frac{f^{(n)}(\mu)}{n!} \quad \mu \in [a, b].$$

This result can be generalized and we can write

$$f(x, x_0, x_1, \ldots, x_n) = \frac{f^{(n+1)}(\mu)}{(n + 1)!} \qquad (2.17)$$

where there are now $(n + 2)$ points $x, x_0, \ldots, x_n$ involved and μ is in the interval $[x, x_0, \ldots, x_n]$. From equations (2.14) and (2.17), it follows that

$$f(x) = p(x) + \pi(x)\frac{f^{(n+1)}(\mu)}{(n + 1)!}, \qquad (2.18)$$

where $\pi(x)$ is given by equation (2.15). Now, although in general we do not know μ, we can frequently find the upper limit of $|f^{n+1}(x)|$ in the interval $[a, b]$ involved. Suppose that $|f^{n+1}(x)| \leqslant K_{n+1}$ for $x \in [a, b]$, then

$$|R(x)| \leqslant K_{n+1}\frac{|\pi(x)|}{(n + 1)!}, \quad (x \in [a, b]). \quad (2.19)$$

The graph of $|\pi(x)|$ is of the form shown in figure 5, where it is assumed that $x_0 < x_1 < \ldots < x_9$. This graph shows that

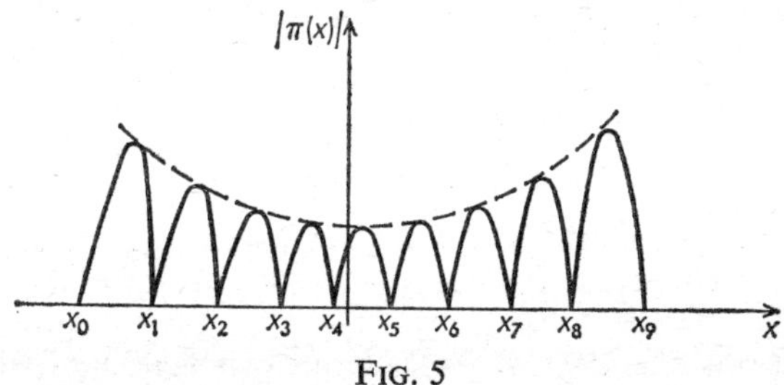

FIG. 5

$|\pi(x)|$ has relative maxima which are smaller in the middle of the range than at the ends. It is therefore desirable to use the interpolation formula in the middle of the range.

Example 13

Given the following data

x	0	$\frac{1}{2}$	2
$f(x) = e^x$	1	1·649	7·389

estimate the value of f(1) and the maximum error.

Solution

It is useful to form the divided difference table:

x	$f(x)$		
0	1		
		1·298	
$\frac{1}{2}$	1·649		1·131
		3·560	
2	7·389		

Using equation (2.14) for p(x), we obtain

$$p(1) = 1 + 1{\cdot}298 + \tfrac{1}{2} \times 1{\cdot}131$$
$$= 2{\cdot}864.$$

This is not a good estimate of $e^1 = 2{\cdot}718$. In the interval $[0, 2]$ equation (2.19) gives the estimate of the maximum error:

$$|R(x)| \leqslant 7{\cdot}4 \frac{|(1 - 0)(1 - \tfrac{1}{2})(1 - 2|}{6} = 0{\cdot}62.$$

Although the estimate of e^1 in the above example was poor, it was well within the bounds of the maximum error for this interpolation.

The maximum error depends on two factors. The first

factor, $f^{(n+1)}(\mu)$; $\mu \in [a, b]$, depends on the function being approximated. The second factor,

$$\frac{1}{(n+1)!} |(x - x_0)(x - x_1) \ldots (x - x_n)|,$$

is independent of the function being interpolated. Although nothing can be done to minimize the first factor except to reduce the interval of interpolation $[a, b]$, the magnitude of the second factor depends on the choice of tabular points $x_0, x_1, \ldots, x_n$. To discuss the best choice of tabular points so as to minimize the contribution to the total error, we would need to consider the theory of Tschebyscheff polynomials. Unfortunately we cannot do this in this small volume, but the interested reader will find this theory adequately outlined in several texts (including reference [6]). Actually we have already met a similar problem in Section 1.6 when we had to find the constant C which minimized the function ε in a specified interval.

Sometimes there are difficulties in evaluating the error term in the form of equation (2.19). For example it is by no means a simple matter to differentiate $(n + 1)$ times $(n \geqslant 1)$ the function

$$f(x) = [(1 + x)^{\frac{1}{2}}(1 - 2x)^{\frac{1}{3}}]^{\frac{1}{2}}.$$

There exist estimates of the error term to overcome this and other similar difficulties. The interested reader should consult reference [5].

2.4 Equally spaced tabular points

The basic property used in the theory of polynomial function interpolation is that a unique polynomial of degree n passes through the $(n + 1)$ points $x_0, x_1, \ldots, x_n$ (for convenience chosen so that $x_0 < x_1 < \ldots < x_n$) at which it takes on the values $f_0, f_1, \ldots, f_n$. This means that, for example, Newton's interpolation formula with finite differences of degree n must be equivalent to Newton's formula with divided differences of the same degree when the $(n + 1)$ tabular points are equally spaced. We will now show this in the case of Newton's forward differences formula. With a little ingenuity it can be shown for other interpolation formulae, i.e. for Bessel's, Everett's, etc.

Since the tabular points are equally spaced

$$x_{i+1} - x_i = h \qquad (i = 0, 1, \ldots, n - 1).$$

Let

$$x = x_0 + ph,$$

so that

$$x - x_i = (p - i)h \qquad (i = 0, 1, \ldots, n - 1).$$

Newton's interpolation formula with divided differences is (equation (2.14))

$$f(x) = f(x_0) + (x - x_0)f(x_0, x_1) +$$
$$+ (x - x_0)(x - x_1)f(x_0, x_1, x_2) + \ldots +$$
$$+ (x - x_0)(x - x_1)\ldots(x - x_{n-1})f(x_0, x_1, \ldots, x_n) +$$
$$+ R(x).$$

Using equations (2.12) we can write the above as

$$f(x_0 + ph) = f(x_0) + ph\,\frac{1}{h}\,\Delta f(x_0) +$$
$$+ ph(p - 1)h\,\frac{1}{2!h^2}\,\Delta^2 f(x_0) + \ldots +$$
$$+ ph(p - 1)h \ldots (p - n + 1)h\,\frac{1}{n!h^n}\,\Delta^n f(x_0) + R(x),$$

that is,

$$f_p = f_0 + \binom{p}{1}\Delta f_0 + \binom{p}{2}\Delta^2 f_0 + \ldots + \binom{p}{n}\Delta^n f_0 + R(x),$$

which is the same as equation (1.13), Newton's interpolation formula including the remainder term.

An interesting observation can be made about the remainder term $R(x)$:

$$R(x) = (x - x_0)(x - x_1)\ldots(x - x_n)f(x, x_0, \ldots, x_n)$$
$$= \binom{p}{n+1}\Delta^{n+1}f_\mu \quad (\mu \in \text{interval } [x, x_0, \ldots, x_n]),$$

which must of course be the same as the expression for $R_{n+1}(p)$ of equation (1.13). In this above form the remainder term is of no particular practical use, but is of theoretical interest. It also ties up with the statements made in Section 1.9 when we were discussing the error in Simpson's rule.

III SIMULTANEOUS LINEAR EQUATIONS

The system of n linear equations in n unknowns,

$$
\begin{aligned}
a_{11}x_1 + a_{12}x_2 + \ldots + a_{1n}x_n &= b_1, \\
a_{21}x_1 + a_{22}x_2 + \ldots + a_{2n}x_n &= b_2, \\
\cdot \quad \cdot \quad \cdot \quad \cdot \quad \cdot \quad \cdot \quad \cdot \quad \cdot \quad & \\
a_{n1}x_1 + a_{n2}x_2 + \ldots + a_{nn}x_n &= b_n,
\end{aligned}
\tag{3.1}
$$

is written in matrix form as:

$$
\mathbf{A}\mathbf{x} = \mathbf{b},
\tag{3.2}
$$

where

$$
\mathbf{A} = \begin{pmatrix}
a_{11} & a_{12} & \ldots & a_{1n} \\
a_{21} & a_{22} & \ldots & a_{2n} \\
\cdot & \cdot & \cdot & \cdot \\
a_{n1} & a_{n2} & \ldots & a_{nn}
\end{pmatrix}
$$

is a matrix of order $n \times n$, and

$$
\mathbf{x} = \begin{pmatrix} x_1 \\ x_2 \\ \cdot \\ \cdot \\ \cdot \\ x_n \end{pmatrix}
\quad \text{and} \quad
\mathbf{b} = \begin{pmatrix} b_1 \\ b_2 \\ \cdot \\ \cdot \\ \cdot \\ b_n \end{pmatrix}
$$

are column vectors of order n.

When the matrix $\mathbf{A}$ is non-singular, that is, when the determinant $|\mathbf{A}|$ is non-zero, the system of equations has a unique solution. A solution consists of an n-tuple of numbers $x_1, x_2, \ldots, x_n$, the components of a vector $\mathbf{x}$, such that

$$
\mathbf{A}\mathbf{x} = \mathbf{b}.
$$

Here we have a classic example where the analytical solution is apparently very simple, i.e.

$$
\mathbf{x} = \mathbf{A}^{-1}\mathbf{b},
$$

but where in practice the work involved in evaluating the inverse matrix $\mathbf{A}^{-1}$ for $n \geqslant 4$ (say) is very tedious. The necessity for numerical techniques which reduce the amount of computation and yet lead to accurate solutions is obvious. There are basically two types of techniques for solving systems of simultaneous equations, direct and iterative.

3.1 Direct methods

The Gaussian elimination method

For reasons which will be explained in Section 3.4 we choose as the *pivot* the equation of the system (3.1) which has the numerically largest coefficient of x_1. This choice is explained in Section 3.5 when we discuss the propagation of errors in calculations.

We add appropriate multiples of the pivot to the $(n - 1)$ remaining equations so that the resulting equations have no term in x_1.

From this new set of $(n - 1)$ equations we choose as pivot the equation having the numerically largest coefficient of x_2. We add appropriate multiples of this new pivot to the remaining $(n - 2)$ equations so that the resulting equations have no term in x_2.

This process is repeated until we end up with a system of equations of the following triangular form.

$$c_{11}x_1 + c_{12}x_2 + \ldots + c_{1n}x_n = d_1,$$
$$c_{22}x_2 + \ldots + c_{2n}x_n = d_2,$$
$$\cdot \quad \cdot \quad \cdot \quad \cdot \quad \cdot \quad \cdot \quad \cdot \quad \cdot \quad \cdot,$$
$$c_{nn}x_n = d_n.$$

In this form the matrix $\mathbf{C}$ of the coefficients c_{ij}, e.g.

$$\mathbf{C} = \begin{pmatrix} c_{11} & c_{12} & c_{13} & \ldots & c_{1n} \\ 0 & c_{22} & c_{23} & \ldots & c_{2n} \\ 0 & 0 & c_{33} & \ldots & c_{3n} \\ \cdot & \cdot & \cdot & & \\ \cdot & \cdot & \cdot & & \\ \cdot & \cdot & \cdot & & \\ 0 & 0 & 0 & \ldots & c_{nn} \end{pmatrix},$$

is called an *upper triangular* matrix.

Two systems of equations $\mathbf{Ax} = \mathbf{b}$ and $\mathbf{Cx} = \mathbf{d}$ having the same solutions are said to be *equivalent*. In fact we could find a linear transformation (i.e. a matrix) which would transform one system into the other (see reference [4]).

We now use *back-substitutions* to solve this system. This means that having determined from the last equation the value of x_n, we substitute it backwards into the last but one equation to determine x_{n-1}. Substituting backwards in this manner we finally find x_1 from the first equation.

When the calculations are being carried out by hand or with the aid of a hand calculating machine it is sensible to have a running check on the calculations. This is done by introducing a 'cross-sum' column, denoted by Σ, whose elements are the sums of all the constant terms (including the right-hand side) of each equation. Each cross-sum is treated in the same way as the equation and should check with the new cross-sum.

Example 14

Use the Gauss elimination method with a running check to solve the following system of equations:

$$4x_1 + x_2 + x_3 = 1,$$
$$-x_1 + 2x_2 - x_3 = 0.5,$$
$$2x_1 - x_2 + 2x_3 = 2.$$

Solution

The first equation of the system serves as a pivot. We write the equations and the Σ column in the following form:

x_1	x_2	x_3	b	Σ
4	1	1	1	7
−1	2	−1	0·5	0·5
2	−1	2	2	5

To eliminate the coefficients of x_1 in the last two equations we must add to them appropriate multiples of the pivot. If we denote these multiples by m_{12} and m_{13} respectively, we must have

$$4m_{12} - 1 = 0,$$

i.e.
$$m_{12} = \tfrac{1}{4}$$

and
$$4m_{13} + 2 = 0,$$

i.e.
$$m_{13} = -\tfrac{1}{2}.$$

The new system of equations is

x_1	x_2	x_3	b	Σ
4	1	1	1	7
	2·25	−0·75	0·75	2·25
	−1·5	1·5	1·5	1·5

Notice how the running check works. For example the second component of the Σ column 2.25 must equal the sum of the elements in the second row. It also must equal

$$7m_{12} + 0·5.$$

If these two results are not equal, some error has been made.

The pivot this time is the second equation and the appropriate multiple m_{23} is found from the relation

$$2·25m_{23} - 1·5 = 0,$$

i.e.
$$m_{23} = \frac{1·5}{2·25} = \tfrac{2}{3}.$$

The new system of equations is

x_1	x_2	x_3	b	Σ
4	1	1	1	7
	2·25	−0·75	0·75	2·25
		1·0	2·0	3·0

Now we use back substitutions to find

$$x_3 = 2,$$

$$2·25x_2 - 0·75x_3 = 0·75, \quad \text{i.e. } x_2 = 1,$$

$$4x_1 + x_2 + x_3 = 1, \quad \text{i.e. } x_1 = -0·5.$$

There are a number of variations of the Gauss elimination method which we shall not discuss here; an interesting discussion will be found in reference [5].

Another direct method for solving systems of simultaneous equations is Choleski's method.

Choleski's method

This method is of great interest since it depends on matrix properties. If we consider the system of simultaneous equation (3.2), i.e.

$$\mathbf{Ax} = \mathbf{b},$$

which is such that no leading principal minor* of $\mathbf{A}$ is zero, then we can find two matrices $\mathbf{L}$ and $\mathbf{U}$ so that

$$\mathbf{A} = \mathbf{LU}. \tag{3.3}$$

The system of equations can then be written as

$$\mathbf{LUx} = \mathbf{b}. \tag{3.4}$$

We choose $\mathbf{L}$ and $\mathbf{U}$ to have rather special forms. If $\mathbf{A}$ is a matrix of order 3×3, then $\mathbf{L}$ has the form

$$\mathbf{L} = \begin{pmatrix} l_{11} & 0 & 0 \\ l_{21} & l_{22} & 0 \\ l_{31} & l_{32} & l_{33} \end{pmatrix}$$

and $\mathbf{U}$ has the form

$$\mathbf{U} = \begin{pmatrix} 1 & u_{12} & u_{13} \\ 0 & 1 & u_{23} \\ 0 & 0 & 1 \end{pmatrix}.$$

For obvious reasons $\mathbf{L}$ is known as a *lower triangular* matrix and $\mathbf{U}$ as an *upper triangular* matrix. We could have chosen $\mathbf{L}$ to have unit elements along its leadin diagonal, but then the elements of the leading diagonal of $\mathbf{U}$, namely $u_{11}, u_{22}, \ldots,$

---

* *This means that the elements of the matrix A are such that*

$$a_{11} \neq 0, \begin{vmatrix} a_{11} & a_{12} \\ a_{21} & a_{22} \end{vmatrix} 0, \ldots, |A| \neq 0.$$

(See reference [7], p. 17).

would, in general, differ from unity. Equation (3.4) has the form

$$\mathbf{Ly} = \mathbf{b},\qquad(3.5)$$

where

$$\mathbf{y} = \mathbf{Ux}.\qquad(3.6)$$

Because of the particular form of the matrix $\mathbf{L}$ the system of equations (3.5) can be solved quite simply to obtain $\mathbf{y}$. Further, because of the particular form of the matrix $\mathbf{U}$, the system of equations (3.6) can be solved by back-substitution.

Example 15

Solve the system of equations of Example 14 by Choleski's method.

$$
\begin{aligned}
4x_1 + x_2 + x_3 &= 1, \\
-x_1 + 2x_2 - x_3 &= 0\cdot 5, \\
2x_1 - x_2 + 2x_3 &= 2.
\end{aligned}
$$

Solution

We write the matrix of the coefficients $\mathbf{A}$ as the product of the two triangular matrices $\mathbf{L}$ and $\mathbf{U}$, i.e.

$$
\begin{pmatrix} 4 & 1 & 1 \\ -1 & 2 & -1 \\ 2 & -1 & 2 \end{pmatrix} = \begin{pmatrix} l_{11} & 0 & 0 \\ l_{21} & l_{22} & 0 \\ l_{31} & l_{32} & l_{33} \end{pmatrix} \begin{pmatrix} 1 & u_{12} & u_{13} \\ 0 & 1 & u_{23} \\ 0 & 0 & 1 \end{pmatrix}.
$$

On equating the corresponding elements of the matrices, we obtain the following nine equations:

$$
\begin{aligned}
4 &= l_{11}, \\
1 &= l_{11}u_{12}, \\
1 &= l_{11}u_{13}, \\
-1 &= l_{21}, \\
2 &= l_{21}u_{12} + l_{22}, \\
-1 &= l_{21}u_{13} + l_{22}u_{23}, \\
2 &= l_{31}, \\
-1 &= l_{31}u_{12} + l_{32}, \\
2 &= l_{31}u_{13} + l_{32}u_{23} + l_{33}.
\end{aligned}
$$

On solving these equations we determine the two triangular matrices, they are

$$\begin{pmatrix} 4 & 0 & 0 \\ -1 & \frac{9}{4} & 0 \\ 2 & -\frac{3}{2} & 1 \end{pmatrix} \text{ and } \begin{pmatrix} 1 & \frac{1}{4} & \frac{1}{4} \\ 0 & 1 & -\frac{1}{3} \\ 0 & 0 & 1 \end{pmatrix}.$$

Next we find $\mathbf{y}$ such that $\mathbf{Ly} = \mathbf{b}$, i.e.

$$\begin{pmatrix} 4 & 0 & 0 \\ -1 & \frac{9}{4} & 0 \\ 2 & -\frac{3}{2} & 1 \end{pmatrix}\begin{pmatrix} y_1 \\ y_2 \\ y_3 \end{pmatrix} = \begin{pmatrix} 1 \\ 0 \cdot 5 \\ 2 \end{pmatrix}.$$

This is simple to solve; we find that

$$\mathbf{y} = \begin{pmatrix} \frac{1}{4} \\ \frac{1}{3} \\ 2 \end{pmatrix}.$$

Finally we have to solve the equation (3.6) which for this example is:

$$\begin{pmatrix} \frac{1}{4} \\ \frac{1}{3} \\ 2 \end{pmatrix} = \begin{pmatrix} 1 & \frac{1}{4} & \frac{1}{4} \\ 0 & 1 & -\frac{1}{3} \\ 0 & 0 & 1 \end{pmatrix}\begin{pmatrix} x_1 \\ x_2 \\ x_3 \end{pmatrix}.$$

We solve this system of equations by back substitution to find

$$x_3 = 2,$$
$$x_2 = 1,$$

and
$$x_1 = -\tfrac{1}{2}.$$

Choleski's method is particularly useful when the matrix of the coefficients of $\mathbf{A}$ is symmetric. In this case

$$\mathbf{A} = \mathbf{LU},$$
$$\mathbf{A}' = \mathbf{U}'\mathbf{L}'$$

(the dash stands for the transposed matrix), and since

$$\mathbf{A} = \mathbf{A}',$$
$$\mathbf{U} = \mathbf{L}'.$$

It follows that $\mathbf{A} = \mathbf{LL}'$ and we need only compute the components of one triangular matrix instead of two. In this case the diagonal components may not be all unity, indeed some of the elements of the matrix may be imaginary.

When carrying out the above calculations repeatedly by hand it is advisable to have a proper lay-out. A final check is always necessary; it is to substitute the calculated values of x_i $(i = 1, 2, \ldots, n)$ into the original equations, but a running check can also be useful. Readers interested in these practical considerations are advised to consult reference [5].

Before discussing iterative methods for solving a system of simultaneous equations we shall consider the problem of inverting a non-singular matrix.

3.2 Computing the inverse of a matrix

It has already been pointed out in the beginning of this chapter that the amount of work involved in finding the reciprocal of a matrix $\mathbf{A}$, that is $\mathbf{A}^{-1}$ so as to find the solution to the system of equation $\mathbf{Ax} = \mathbf{b}$, is rather large. But sometimes it is nevertheless desirable to find $\mathbf{A}^{-1}$. For example, we may need to solve systems of equations

$$\mathbf{Ax} = \mathbf{b}$$

for which only the right-hand sides differ. The methods we use to compute $\mathbf{A}^{-1}$ are paradoxically enough the same ones we use to solve a system of simultaneous equations.

We consider for simplicity the system of equations

$$a_{11}x_1 + a_{12}x_2 = b_1,$$
$$a_{21}x_1 + a_{22}x_2 = b_2. \tag{3.7}$$

We assume that the inverse of the non-singular matrix

$$\mathbf{A} = \begin{pmatrix} a_{11} & a_{12} \\ a_{21} & a_{22} \end{pmatrix}$$

is the matrix

$$\mathbf{A}^{-1} = \begin{pmatrix} \alpha_{11} & \alpha_{12} \\ \alpha_{21} & \alpha_{22} \end{pmatrix}.$$

It follows that the solution to the system of equations (3.7) is

$$\begin{pmatrix} x_1 \\ x_2 \end{pmatrix} = \begin{pmatrix} \alpha_{11} & \alpha_{12} \\ \alpha_{21} & \alpha_{22} \end{pmatrix} \begin{pmatrix} b_1 \\ b_2 \end{pmatrix}. \tag{3.8}$$

If we choose $\mathbf{b} = \begin{pmatrix} 1 \\ 0 \end{pmatrix}$ then from equation (3.8) we see that the solution to the system (3.7) is

$$\begin{pmatrix} x_1 \\ x_2 \end{pmatrix} = \begin{pmatrix} \alpha_{11} \\ \alpha_{21} \end{pmatrix}.$$

So by this particular choice of the vector $\mathbf{b}$, the components of the solution vector are the elements of the first column of the inverse matrix $\mathbf{A}^{-1}$.

If we next choose $\mathbf{b} = \begin{pmatrix} 0 \\ 1 \end{pmatrix}$, we find the solution vector

$$\begin{pmatrix} x_1 \\ x_2 \end{pmatrix} = \begin{pmatrix} \alpha_{12} \\ \alpha_{22} \end{pmatrix}.$$

This is the second column of the inverse matrix.

We can generalize these ideas to any non-singular matrix $\mathbf{A}$ of order $n \times n$. To find the inverse matrix $\mathbf{A}^{-1}$ we must solve the n systems of n simultaneous equations

$$\mathbf{Ax} = \mathbf{I} \qquad\qquad (3.8)$$

where the right-hand sides of each system are the column vectors making up the unit matrix $\mathbf{I}$, i.e. the right-hand sides are the unit (or elementary) vectors

$$\mathbf{e}_1 = \begin{pmatrix} 1 \\ 0 \\ 0 \\ \cdot \\ \cdot \\ \cdot \\ 0 \end{pmatrix} \quad \mathbf{e}_2 = \begin{pmatrix} 0 \\ 1 \\ 0 \\ \cdot \\ \cdot \\ \cdot \\ 0 \end{pmatrix} \ldots \mathbf{e}_n = \begin{pmatrix} 0 \\ 0 \\ 0 \\ \cdot \\ \cdot \\ \cdot \\ 1 \end{pmatrix}.$$

In fact, instead of solving each of the above systems separately, we solve them all simultaneously.

We can do this with the aid of any method used for solving systems of simultaneous equations. For example, with the aid of the Gauss elimination method we reduce the matrix of the coefficients $\mathbf{A}$ to an upper triangular matrix by operations already discussed in Section 3.1, and simultaneously we perform the same operations on the n right-hand sides, e_1, e_2, . . ., e_n. We then carry out n separate back-substitutions each resulting in one column of the inverse matrix A^{-1}. This method is illustrated in the example below. Note that we have the same running checks as in Example 14, although in Example 16 below we are dealing with three systems of equations.

Example 16

Find the inverse of the matrix of coefficients of the following system of equations:

$$4x_1 + x_2 + x_3 = 1,$$
$$-x_1 + 2x_2 - x_2 = 0{\cdot}5,$$
$$2x_1 - x_2 + 2x_3 = 2.$$

Solution

Since we have already solved this system of equations (Examples 14 and 15), there is no need of further discussion about the choice of pivots, etc. The operations being executed are described in the first column of the worksheet below.

Operation	x_1	x_2	x_3	I			Σ	Eqn.
PIVOT $\Longrightarrow$	4	1	1	1	0	0	7	①
	-1	2	-1	0	1	0	1	②
	2	-1	2	0	0	1	4	③
	4	1	1	1	0	0	7	
② $+ \frac{1}{4}$① PIVOT $\Longrightarrow$	0	$\frac{9}{4}$	$-\frac{3}{4}$	$\frac{1}{4}$	1	0	$1\frac{1}{4}$	④
③ $+ \frac{1}{2}$①	0	$-\frac{3}{2}$	$\frac{3}{2}$	$-\frac{1}{2}$	0	1	$\frac{1}{2}$	⑤
	4	1	1	1	0	0	7	
	0	$\frac{9}{4}$	$-\frac{3}{4}$	$\frac{1}{4}$	1	0	$1\frac{1}{4}$	
⑤ $+ \frac{2}{3}$④	0	0	1	$-\frac{1}{3}$	$\frac{2}{3}$	1	$\frac{7}{3}$	

Having reduced the matrix of coefficients to the upper triangular matrix

$$\begin{pmatrix} 4 & 1 & 1 \\ 0 & \tfrac{9}{4} & -\tfrac{3}{4} \\ 0 & 0 & 1 \end{pmatrix},$$

we must now solve the corresponding systems of equations with the following three right-hand sides:

$$\mathbf{b}_1 = \begin{pmatrix} 1 \\ \tfrac{1}{4} \\ -\tfrac{1}{3} \end{pmatrix}$$

$$\mathbf{b}_2 = \begin{pmatrix} 0 \\ 1 \\ \tfrac{2}{3} \end{pmatrix}$$

and
$$\mathbf{b}_3 = \begin{pmatrix} 0 \\ 0 \\ 1 \end{pmatrix}.$$

The solutions are respectively:

(1) $x_3 = -\tfrac{1}{3},\quad x_2 = 0,\quad x_1 = \tfrac{1}{3},$

(2) $x_3 = \tfrac{2}{3},\quad x_2 = \tfrac{2}{3},\quad x_1 = -\tfrac{1}{3},$

(3) $x_3 = 1,\quad x_2 = \tfrac{1}{3},\quad x_1 = -\tfrac{1}{3}.$

These solutions are the three columns of the inverse matrix

$$\mathbf{A}^{-1} = \begin{pmatrix} \tfrac{1}{3} & -\tfrac{1}{3} & -\tfrac{1}{3} \\ 0 & \tfrac{2}{3} & \tfrac{1}{3} \\ -\tfrac{1}{3} & \tfrac{2}{3} & 1 \end{pmatrix}.$$

This can be checked by multiplication of $\mathbf{A}$ and $\mathbf{A}^{-1}$,

$$\mathbf{A}\mathbf{A}^{-1} = \mathbf{I}.$$

As already mentioned, although only two direct methods for solving systems of simultaneous equations have been discussed in this book, there are other methods which may be more appropriate for certain problems involving special types of matrices. There are a number of difficulties which may arise in solving systems of equations by the two methods discussed which have not been considered in this book. For example, using Choleski's method it may happen that even when the matrix of coefficients **A** is non-singular, one of the diagonal terms of the triangular matrices (**L** or **U**) may be zero. If this happens, we could not of course determine some of the elements of the triangular matrices since they involve a division by this diagonal element. For a discussion of problems such as this, the reader is recommended to consult a more advanced text, for example, reference [5]. In evaluating the relative efficiency of the two methods discussed it is useful to consider among other factors the number of arithmetical operations (i.e. multiplications and additions) involved. Curiously enough, it can be shown (see reference [5]) that for a system of n equations in n unknown both methods involve the same number of operations

$$\tfrac{1}{3}n^3 + n^2 - \tfrac{1}{3}n \text{ multiplications,}$$

and

$$\tfrac{1}{3}n^3 + \tfrac{1}{2}n^2 - \tfrac{5}{6}n \text{ additions.}$$

3.3 Iterative methods

The philosophy behind iterative methods is quite different to the one for direct methods. In the case of iterative methods we start with either an arbitrary approximation or a first crude approximation to the solution, which is then improved until some specified accuracy is achieved. These approximations are obtained by iterative procedures which are ideally suited for computer work. Clearly, the issue of convergence of the iterations is very important. Even when the iterations do converge to the solution, the convergence may be slow, and it may often be more efficient to use a direct method. An iterative method is particularly useful for a system of equations whose matrix of coefficients is 'sparse', that is, a matrix which has many zero coefficients. Such matrices arise in certain control problems and in the numerical solution of partial differential equations.

The Jacobi method

We consider the system of equations

$$a_{11}x_1 + a_{12}x_2 + a_{13}x_3 = b_1,$$
$$a_{21}x_1 + a_{22}x_2 + a_{23}x_3 = b_2, \qquad (3.9)$$
$$a_{31}x_1 + a_{32}x_2 + a_{33}x_3 = b_3.$$

We start with an arbitrary approximation to the solution, say

$$x_1^{(0)} = x_2^{(0)} = x_3^{(0)} = 1.$$

Substituting these values into the equations (3.9) we obtain new approximations $x_1^{(1)}$, $x_2^{(1)}$, $x_3^{(1)}$ to the solutions. This process is repeated until sufficient accuracy is obtained. The iterative procedure is described by the equations

$$x_1^{(r+1)} = \frac{1}{a_{11}} (b_1 - a_{12}x_2^{(r)} - a_{13}x_3^{(r)}),$$

$$x_2^{(r+1)} = \frac{1}{a_{22}} (b_2 - a_{21}x_1^{(r)} - a_{23}x_3^{(r)}), \quad (3.10)$$

$$x_3^{(r+1)} = \frac{1}{a_{33}} (b_3 - a_{31}x_1^{(r)} - a_{32}x_2^{(r)}).$$

We can describe the above procedure in matrix notation. Let

$$\mathbf{L} = \begin{pmatrix} 0 & 0 & 0 \\ a_{21} & 0 & 0 \\ a_{31} & a_{32} & 0 \end{pmatrix}$$

$$\mathbf{D} = \begin{pmatrix} a_{11} & 0 & 0 \\ 0 & a_{22} & 0 \\ 0 & 0 & a_{33} \end{pmatrix}$$

and

$$\mathbf{U} = \begin{pmatrix} 0 & a_{12} & a_{13} \\ 0 & 0 & a_{23} \\ 0 & 0 & 0 \end{pmatrix}.$$

So that the matrix of coefficients $\mathbf{A} = \mathbf{L} + \mathbf{D} + \mathbf{U}$. It is simple to verify the that iterative method is represented by the matrix equation

$$\mathbf{D}\mathbf{x}^{(r+1)} = \mathbf{b} - (\mathbf{L} + \mathbf{U})\mathbf{x}^{(r)} \quad (r = 0, 1, 2, \ldots),$$

so that the $(r+1)$th approximation to the solution of the systems of equations (3.10) is

$$\mathbf{x}^{(r+1)} = \mathbf{D}^{-1}\mathbf{b} - \mathbf{D}^{-1}(\mathbf{L} + \mathbf{U})\mathbf{x}^{(r)}, \qquad (3.11)$$

where

$$\mathbf{b} = \begin{pmatrix} b_1 \\ b_2 \\ b_3 \end{pmatrix}$$

and

$$\mathbf{x}^{(r)} = \begin{pmatrix} x_1^{(r)} \\ x_2^{(r)} \\ x_3^{(r)} \end{pmatrix}.$$

Example 17

Use the Jacobi method to solve the following system of simultaneous equation:

$$2x_1 - x_2 = 0,$$
$$x_1 + 2x_2 = 5.$$

Solution

By inspection it is seen that the solution is

$$\mathbf{x} = \begin{pmatrix} 1 \\ 2 \end{pmatrix}.$$

The iterations are defined by:

$$x_1^{(r+1)} = \tfrac{1}{2}x_2^{(r)},$$
$$x_2^{(r+1)} = \tfrac{1}{2}[5 - x_1^{(r)}].$$

Beginning with $x_1^{(0)} = x_2^{(0)} = 1$, the next two approximations are

$$\mathbf{x}^{(1)} = \begin{pmatrix} \tfrac{1}{2} \\ 2 \end{pmatrix} \quad \text{and} \quad \mathbf{x}^{(2)} = \begin{pmatrix} 1 \\ 2 \end{pmatrix}.$$

The sequence of approximations is seen to converge very rapidly to the solution.

In this example

$$\mathbf{L} = \begin{pmatrix} 0 & 0 \\ 1 & 0 \end{pmatrix},$$

$$D = \begin{pmatrix} 2 & 0 \\ 0 & 2 \end{pmatrix},$$

and

$$U = \begin{pmatrix} 0 & -1 \\ 0 & 0 \end{pmatrix}.$$

Note that in both rows of A the diagonal elements (the non-zero elements of D) are large relative to the other elements.

Convergence

We define the *error vector* of the rth iteration by

$$e^{(r)} = x - x^{(r)}, \tag{3.12}$$

where x is the solution vector and $x^{(r)}$ is the rth approximation to it.

Since

$$(L + D + U)x = b,$$
$$x = D^{-1}b - D^{-1}(L + U)x.$$

From the above equation and equation (3.11) we have

$$\begin{aligned} e^{(r+1)} &= x - x^{(r+1)} \\ &= -D^{-1}(L + U)(x - x^{(r)}) \\ &= -D^{-1}(L + U)e^{(r)}. \end{aligned}$$

Hence

$$e^{(r+1)} = \{-D^{-1}(L + U)\}^{r+1}e^{(0)}.$$

The vector $e^{(0)}$ is a constant vector; it is the difference between the solution vector and the vector having unit elements. It follows that the convergence of the method depends on the behaviour of $\{-D^{-1}(L + U)\}^n$ as n increases.

If we let

$$\{-D^{-1}(L + U)\} = B,$$

the convergence of the method is assured if

$$\lim_{r \to \infty} e^{(r)} = 0, \text{ i.e. if } B^r \longrightarrow O.$$

A matrix is said to approach zero if all its elements converge to zero. To find the conditions necessary for this to happen for the Jacobi method and for other methods (the Gauss–Seidel method, for example), we need to introduce and study *norms* of vectors and matrices. Norms are numbers indicating the 'size' of a particular vector or matrix. A well-known norm is the Euclidean one which associates the number $\left[\sum_{i=1}^{n} x_i^2\right]^{\frac{1}{2}}$ with the vector where components are $x_1, x_2, \ldots, x_n$. Unfortunately we do not have the space in this book to develop the theory of norms, but a good discussion of it can be found in references [8] and [9].

The conclusion of such an investigation results in the following sufficient condition for the convergence of the iterative technique:

$$\max_i \left\{ \frac{1}{|a_{ij}|} \sum_{\substack{j=1 \\ j \neq i}}^{n} |a_{ij}| \right\} < 1 \quad (i, j = 1, 2, \ldots, n). \quad (3.13)$$

The a_{ij}'s are the elements of the matrix of coefficients, **A**. Equation (3.13) implies that for each row of the matrix the sum of the moduli of the off-diagonal elements must be smaller than the modulus of the diagonal element. It is interesting to note that this sufficient condition (equation 3.13) is independent of the arbitrary initial approximation $\mathbf{x}^{(0)}$ which was chosen to be the vector of unit elements.

Although we have a sufficient condition for convergence, it does not guarantee *rapid* convergence. This is rather important because unless the convergence is rapid it may be much better to solve the system of equation by direct methods. There is a fairly simple empirical rule (see reference [9]) which gives some indication of the speed of convergence.

Let S_i = sum of the moduli of the non-diagonal elements, in row i ($i = 1, 2, \ldots, n$). Thus,

$$S_i = |a_{i1}| + |a_{i2}| + \ldots + |a_{i,i-1}| + |a_{i,i+1}| + \ldots + |a_{in}|$$

of the matrix of coefficients **A** and

$$q_i = \frac{S_i}{|a_{ii}| - S_i},$$

where a_{ii} is the diagonal element in row i. Further, let

$$q = \max_i q_i \quad (i = 1, 2, \ldots, n).$$

It is obvious from equation (3.13) that for convergence we must have

$$q_i > 0 \quad (i = 1, 2, \ldots, n).$$

If $q < \varepsilon < 2$ the convergence is rapid, and all the more rapid, the smaller ε is.

The Gauss–Seidel method

It is assumed that the system of equations (3.9) has been rearranged so that a_{ii} is the largest coefficient in the ith equation ($i = 1, 2, 3$).

In the first equation we let $x_2 = x_3 = 0$ and determine x_1. This value of x_1 is substituted in the second equation from which x_2 is determined, still assuming that $x_3 = 0$.

Finally these values of x_1 and x_2 are substituted in the third equation to determine x_3. This process is now repeated, using the new approximations to the solutions. This iterative procedure may be described in the following way:

$$x_1^{(r+1)} = \frac{1}{a_{11}} (b_1 - a_{12}x_2^{(r)} - a_{13}x_3^{(r)}),$$

$$x_2^{(r+1)} = \frac{1}{a_{22}} (b_2 - a_{21}x_1^{(r+1)} - a_{23}x_3^{(r)}), \quad (3.14)$$

$$x_3^{(r+1)} = \frac{1}{a_{33}} (b_3 - a_{31}x_1^{(r+1)} - a_{32}x_2^{(r+1)})$$

$$(r = 0, 1, 2, \ldots).$$

Using the matrix notation already defined, the Gauss–Seidel method is described by the equation:

$$(\mathbf{D} + \mathbf{L})\mathbf{x}^{(r+1)} = \mathbf{b} - \mathbf{U}\mathbf{x}^{(r)} \quad (r = 0, 1, 2, \ldots).$$

Corresponding to the system of equation (3.9), we have the $(r + 1)$th approximation to the solution in matrix form:

$$\mathbf{x}^{(r+1)} = (\mathbf{D} + \mathbf{L})^{-1}\mathbf{b} - (\mathbf{D} + \mathbf{L})^{-1}\mathbf{U}\mathbf{x}^{(r)}. \quad (3.15)$$

Note the difference in the matrix equation (3.11) and (3.15) characterizing the Jacobi and the Gauss–Seidel methods respectively.

Example 18

Use the Gauss–Seidel method to solve the system of equations:

$$10x_1 - x_2 = 11,$$
$$x_1 + 5x_2 = -4.$$

Solution

By inspection it is seen that the solution is

$$x_1 = 1, \quad x_2 = -1.$$

The iterations are defined by:

$$x_1^{(r+1)} = \tfrac{1}{10}(11 + x_2^{(r)}),$$
$$x_2^{(r+1)} = \tfrac{1}{5}(-4 - x_1^{(r+1)}).$$

The sequence of approximations converges rapidly. Indeed the first two approximations are

$$\mathbf{x}^{(1)} = \begin{pmatrix} \tfrac{11}{10} \\ -\tfrac{51}{50} \end{pmatrix} \quad \text{and} \quad \mathbf{x}^{(2)} = \begin{pmatrix} \tfrac{499}{500} \\ -\tfrac{2499}{2500} \end{pmatrix}.$$

Convergence

The convergence considerations of the Gauss–Seidel method are similar to the ones for the Jacobi method. Although at first sight it may appear that the Gauss–Seidel method is more likely to converge than the Jacobi method, this is not true in general. In fact, in reference [5] there is an example of a system for which the Jacobi method converges, whereas the Gauss–Seidel method fails to converge.

3.4 Rounding errors and ill-conditioning

When solving systems of simultaneous equations, we have up till now considered coefficients which were exact numbers. We shall now consider the possible effect of rounding errors on the solutions. These are unavoidable when using either a hand calculating machine or a computer.

Example 19

Solve the system of equations

$$2 \cdot 5x_1 + 3x_2 = b_1,$$
$$-2 \cdot 1x_1 - 2 \cdot 5x_2 = b_2,$$

(a) when $b_1 = -3 \cdot 51$ and $b_2 = 2 \cdot 88$, and

(b) when the right hand sides are rounded off to one decimal place, i.e. $b_1 = -3 \cdot 5$ and $b_2 = 2 \cdot 9$.

Solution

Using one of the direct methods discussed in section 3.1, we find that the solutions are

(a) $x_1 = -2 \cdot 7$ and $x_2 = -3 \cdot 42$,

(b) $x_1 = 1$ and $x_2 = -2$.

Note that the two sets of solutions are quite different. Systems such as the one above for which small changes in the data results in large changes in the solutions are said to be *ill-conditioned*. In example 19 we had only considered small changes in the right-hand sides of the equations, but similar large changes in the solutions are obtained for small changes in the coefficients of the equations or indeed for small changes in both the coefficients and the right-hand sides.

Although admittedly examples similar to the above are extreme ones, nevertheless they do underline the importance of making an assessment of errors in calculations, this is especially important in cases of ill-conditioned systems. In the example above we can explain why the system is ill-conditioned in terms of simple geometrical concepts. The two equations represent two lines which are very nearly parallel. The solution to the system, say (x_1, x_2), comprises the coordinates of the point P of intersection of the two lines. Any small change in the data will obviously have a very marked effect on the position of P and hence on the solution. The fact that the two lines are nearly parallel means that the two columns of the matrix of coefficients are almost linearly dependent, so that the matrix is nearly singular. The terms 'almost' linearly dependent and 'nearly' singular are not precise, but they give at least an intuitive explanation of what is happening.

$$P(x_1, x_2)$$

FIG. 6

An example of a 'nearly' singular matrix is

$$\begin{pmatrix} 1 & 1 \\ 1 & 1 + \varepsilon \end{pmatrix},$$

where ε is small. A very small change in ε could have a very marked effect on the solution of the system having the above as the matrix of its coefficients.

If we had a system consisting of several simultaneous equations in more than (say) two or three variables, we could no longer explain ill-conditioning by simple geometric interpretations. But it is still true that in such cases the matrix of coefficients is 'nearly' singular.

Errors

In Section 1.4 we investigated the propagation of errors through a difference table. We shall now consider the propagation of errors in a calculation.

If X is the exact number and x is an approximation to it, the *absolute error* ε is defined as

$$\varepsilon = |X - x|.$$

The (absolute) *relative error* is defined as

$$\frac{|\text{exact number} - \text{approximation}|}{|\text{exact number}|} = \frac{\varepsilon}{|X|}.$$

Usually we are given the approximation x and do not

know X (except within the error limits). The relative error is then approximated by

$$\frac{\varepsilon}{|x|}.$$

Since in general we are interested in the upper bounds of the errors involved, this approximation has no effect on the error analysis so long as ε is small relative to $|x|$. It has already been mentioned that *rounding off* a number to n decimal places involves deleting all digits after the nth digit on the right of the decimal point. If the deleted digits represent a number $\geqslant \frac{1}{2} \times 10^{-n}$, the nth digit is increased by unity, otherwise no change is necessary. This means that the *maximum absolute error* in a number rounded to n decimal places is $\frac{1}{2} \times 10^{-n}$. For example, the number 3·141 592 6 when rounded to six decimal places becomes 3·141 593, and when rounded to two decimal places becomes 3·14. The maximum absolute error in 3·14 is 0·005.

Except for a possible sequence of zeros used to fix the decimal point, the digits in a number are known as *significant figures*. The numbers 432, 0·123, 0·000 789, 6·78 and 100 all have three significant figures.

In *floating point* notation used in computer work every number is written in two parts, the *mantissa* and the *exponent*. The mantissa is always a number smaller than one in modulus. For example, we would write the numbers 123·45 and 0·0067 as $0·123\ 45 \times 10^3$ and $0·67 \times 10^{-2}$. The mantissas are respectively 0·123 45 and 0·67 and the exponents 10^3 and 10^{-2} (or just 3 and -2).

The two numbers $0·12 \times 10^4$ and $0·120 \times 10^4$ have respectively two and three significant figures. The implication of this is that in the first case the maximum absolute error is 50, whereas in the second case it is only 5. When multiplying two numbers we obtain a number which has, in general, more digits than the numbers being multiplied. The question arises: if the two component numbers are approximate, how meaningful are all the digits of the product? This leads to the concept of the *correct* number of significant digits. When an approximate number is rounded off after the nth digit, n being chosen so that the maximum absolute error is $\frac{1}{2} \times 10^{-n}$, we say that the resulting number is correct to n significant figures. For example 0·333 408 is an approximation correct to three significant figures of $\frac{1}{3}$.

3.5 Accumulation of errors in addition (and subtraction)

If x_1 and x_2 are approximations to the numbers X_1 and X_2 with corresponding maximum absolute errors ε_1 and ε_2, then

$$(x_1 - \varepsilon_1) + (x_2 - \varepsilon_2) \leqslant X_1 + X_2 \leqslant (x_1 + \varepsilon_1) + (x_2 + \varepsilon_2),$$

so that

$$(x_1 + x_2) - (\varepsilon_1 + \varepsilon_2) \leqslant X_1 + X_2 \leqslant (x_1 + x_2) + (\varepsilon_1 + \varepsilon_2),$$

i.e.

$$|(X_1 + X_2) - (x_1 + x_2)| \leqslant (\varepsilon_1 + \varepsilon_2).$$

Hence the maximum absolute error of the sum of two numbers is the sum of the absolute errors of the two component numbers. A similar result holds for the difference of two numbers. This result can be generalized in an obvious manner to the sum of n numbers.

Example 20

Find the sum of the three numbers 122·4, 31·25 and 4·175, all correct to four significant figures. Discuss the significance of the number of digits in the answer.

Solution

The sum is 157·825.

Since the maximum absolute errors in the three numbers are 0·05, 0·005 and 0·0005 respectively, the absolute maximum error in the sum is 0·0555. Hence the correct value of the sum is some number in the interval [157·7695, 157·8805]. This shows that the last three digits of the sum are not significant and it is correct to at most four significant figures—in fact the fourth digit '8' is suspect.

A very real problem in numerical calculations arises when

subtraction of two nearly equal numbers takes place. On a computer, when a fixed number of digits are used for the calculations, the loss of accuracy can be such that the answer is meaningless. A possible solution to such a problem is to work with a double accuracy—that is, using twice as many digits in the calculations as is usually done, or else to look again at the calculation being performed and try to reformulate it so as to avoid this particular subtraction.

3.6 Accumulation of errors in multiplication (and division)

Let x_1 and x_2 be approximations to the numbers X_1 and X_2, ε_1 and ε_2 and the corresponding maximum absolute errors.

Then

$$(x_1 - \varepsilon_1)(x_2 - \varepsilon_2) \leqslant X_1 X_2 \leqslant (x_1 + \varepsilon_1)(x_2 + \varepsilon_2),$$

or

$$x_1 x_2 - x_1 \varepsilon_2 - x_2 \varepsilon_1 + \varepsilon_1 \varepsilon_2 \leqslant X_1 X_2 \leqslant x_1 x_2 + x_1 \varepsilon_2 + x_2 \varepsilon_1 + \varepsilon_1 \varepsilon_2.$$

For small ε_1 and ε_2 (relative to X_1 and X_2) we can neglect the product $\varepsilon_1 \varepsilon_2$ in the above expression, hence

$$|X_1 X_2 - x_1 x_2| \leqslant x_1 \varepsilon_2 + x_2 \varepsilon_1,$$

that is,

$$\left| \frac{X_1 X_2 - x_1 x_2}{x_1 x_2} \right| \leqslant \left| \frac{\varepsilon_2}{x_2} \right| + \left| \frac{\varepsilon_1}{x_1} \right|.$$

The above result can be easily generalized to the product of any finite number of approximate numbers and can be stated as follows:

The maximum absolute relative error of the product of n numbers is the sum of the maxima relative errors of the numbers.

A special case of interest in the solution of simultaneous equations is the propagation of error in the multiplication of an approximate number by a fixed number, say m. We find

$$m(x + \varepsilon) = mx + m\varepsilon,$$

so that the absolute error in the result is m times the absolute error in the approximate number.

This is one of the reasons why, in the Gauss elimination method, we choose as pivot the equation having the numerically largest coefficient of x_i $(i = 1, 2, \ldots, n)$. This choice ensures that the multipliers m_{ij} are not greater than 1 in modulus and hence that the absolute errors in the coefficients of the equivalent system of equations are small. A detailed discussion of the propagation of errors in the solution of system of simultaneous equations can be found in references [1] and [4].

By a similar process to the above we can find the absolute relative error in division, indeed

$$\frac{x_1 - \varepsilon_1}{x_2 + \varepsilon_2} \leqslant \frac{X_1}{X_2} \leqslant \frac{x_1 + \varepsilon_1}{x_2 - \varepsilon_2}.$$

On multiplying the numerator and the denominator of the right-hand side by $(x_2 + \varepsilon_2)$ and of the left hand side by $(x_2 - \varepsilon_2)$, we find:

$$\frac{x_1x_2 - x_1\varepsilon_2 - x_2\varepsilon_1 + \varepsilon_1\varepsilon_2}{x_2^2 - \varepsilon_2^2} \leqslant \frac{X_1}{X_2} \leqslant \frac{x_1x_2 + x_1\varepsilon_2 + x_2\varepsilon_1 + \varepsilon_1\varepsilon_2}{x_2^2 - \varepsilon_2^2}.$$

Assuming small errors, we can neglect $\varepsilon_1\varepsilon_2$ and ε_2^2, relative to the other terms, so that the above expression becomes

$$\frac{x_1}{x_2} - \frac{x_1}{x_2}\left(\frac{\varepsilon_2}{x_2} + \frac{\varepsilon_1}{x_1}\right) \leqslant \frac{X_1}{X_2} \leqslant \frac{x_1}{x_2} + \frac{x_1}{x_2}\left(\frac{\varepsilon_2}{x_2} + \frac{\varepsilon_1}{x_1}\right).$$

Hence,

$$\left|\frac{X_1}{X_2} - \frac{x_1}{x_2}\right| \leqslant \left|\frac{x_1}{x_2}\right|\left(\left|\frac{\varepsilon_2}{x_2}\right| + \left|\frac{\varepsilon_1}{x_1}\right|\right).$$

Dividing the above inequality by $\left|\dfrac{x_1}{x_2}\right|$ we obtain the following result:

The maximum absolute relative error of a quotient of two numbers is the sum of the maxima absolute relative errors of the numbers.

Example 21

Discuss the error in the product of the two numbers 123·4 and 0·005 16, assuming each accurate to the given number of significant figures.

Solution

The product is 0·636 744.

The maximum relative error in the two numbers are

$$\frac{0\cdot05}{123\cdot4} = 0\cdot04\% \quad \text{and} \quad \frac{0\cdot000\ 005}{0\cdot005\ 16} = 0\cdot1\% \text{ respectively.}$$

The maximum relative error in the product is therefore

$$0\cdot04 + 0\cdot1 = 0\cdot14\%,$$

i.e. about 1 part in 600. So the product may be in error by one unit in its third significant digit, and we can only be sure of the first two digits being correct.

From the above example we can see that the number of correct significant figures in the product is equal to or less than the number of correct significant figures in the least accurate of the numbers being multiplied.

It must be remembered that in the above analysis we considered the maximum possible effect of errors by considering their maximum absolute values. In many calculations the rounding errors may be of opposite signs and smaller than a half unit in the last decimal place, so that the final solution may be much more accurate than the above analysis suggests. On the other hand in some calculations, for example those involving rounded numbers being raised to a power, the errors are of like sign and the worst possible case of error accumulation may occur.

3.7 Errors in solving systems of simultaneous equations

A rigorous discussion of errors involved in solving systems of simultaneous equations is beyond the scope of this book and the interested reader should examine references [1] and [4]. But we shall briefly discuss the problem of finding the error bounds for the solutions.

94

Let $\mathbf{x}$ be the solution to the system of equations $\mathbf{Ax} = \mathbf{b}$.

It will be assumed that the calculations involved in finding $\mathbf{x}$ are exact and that the only errors involved are due to rounding off the elements a_{ij} of $\mathbf{A}$ and b_i of $\mathbf{b}$. This assumption is of course a gross simplification of the true state of affairs.

We can assume that, before rounding, the element a_{ij} was $a_{ij} + e_{ij}$ and the element b_i was $b_i + d_i$, so that the exact system of equations is

$$(\mathbf{A} + \mathbf{E})\mathbf{x} = \mathbf{b} + \mathbf{d}$$

where the matrix $\mathbf{E}$ has elements e_{ij} and the vector $\mathbf{d}$ has components d_i. We shall assume that the solution to the exact system of equations is $\underline{\mathbf{x}} + \underline{\mathbf{y}}$ and we wish to find the upper bounds for components of $\underline{\mathbf{y}}$.

Since $\underline{\mathbf{x}} + \underline{\mathbf{y}}$ is the solution to the system, we have

$$(\mathbf{A} + \mathbf{E})(\underline{\mathbf{x}} + \underline{\mathbf{y}}) = \mathbf{b} + \mathbf{d}.$$

By assumption both the components of $\mathbf{E}$ and $\mathbf{y}$ are small so that in the above expression we can neglect the contribution of the vector $\mathbf{E}\underline{\mathbf{y}}$. Let $[\mathbf{A}]$ be the matrix whose elements are the moduli of the corresponding elements of $\mathbf{A}$, then

$$[\mathbf{A}][\underline{\mathbf{y}}] \leqslant [\mathbf{d}] + [\mathbf{E}][\underline{\mathbf{x}}]$$

i.e.

$$[\underline{\mathbf{y}}] \leqslant [\mathbf{A}^{-1}][\mathbf{d}] + [\mathbf{A}^{-1}][\mathbf{E}][\underline{\mathbf{x}}].$$

We can now see the difficulties associated with this error analysis, even with the simplifying assumptions we have made. To find $[\underline{\mathbf{y}}]$ we must calculate $\mathbf{A}^{-1}$. This involves as much work as solving the original system of equations. Also it is unlikely that $\mathbf{A}^{-1}$ can be found, without some associated errors in the calculations. So strictly speaking we would need to investigate the effect of the errors in $\mathbf{A}^{-1}$ on our error analysis. Obviously there is no point in going to these lengths since we are only interested in estimating the upper bounds of the components of $[\underline{\mathbf{y}}]$. Unless we employ statistical methods to find the average errors involved, we overestimate the effect of rounding off, since to determine $\mathbf{E}$

and **y** we use the moduli of the maximum absolute errors in rounding the coefficients of **A** and **b**.

Example 22

The right-hand sides and the coefficients in the system of equations

$$2{\cdot}50x_1 + 3{\cdot}00x_2 = -3{\cdot}51,$$
$$-2{\cdot}10x_1 - 2{\cdot}50x_2 = 2{\cdot}88,$$

are correct to two decimal places. Discuss the accuracy of the solution, assuming that the calculations are exact.

Solution

Considering the maximum absolute round-off errors we find

$$[\mathbf{E}] = \begin{pmatrix} 0{\cdot}005 & 0{\cdot}005 \\ 0{\cdot}005 & 0{\cdot}005 \end{pmatrix} \quad \text{and} \quad [\mathbf{d}] = \begin{pmatrix} 0{\cdot}005 \\ 0{\cdot}005 \end{pmatrix}.$$

We have already solved this system of equations (see Example 19, page 88) and found the solutions $x_1 = -2{\cdot}70$ and $x_2 = -3{\cdot}42$. Using one of the methods of inverting matrices discussed in Section 3.2, we find

$$\mathbf{A}^{-1} = \begin{pmatrix} -50 & -60 \\ 42 & 50 \end{pmatrix}.$$

The elements of $\mathbf{A}^{-1}$ are seen to be large relative to the elements of **A**. Indeed this is one characteristic of ill-conditioned systems (see Example 19). From the discussion above it is clear that for ill-conditioned systems the round-off errors in the elements of **A** and **b** will have a relatively large effect on the accuracy of the solution.

For the above example

$$[\mathbf{y}] \leqslant \begin{pmatrix} 50 & 60 \\ 42 & 50 \end{pmatrix}\begin{pmatrix} 0{\cdot}005 \\ 0{\cdot}005 \end{pmatrix} + \begin{pmatrix} 50 & 60 \\ 42 & 50 \end{pmatrix}\begin{pmatrix} 0{\cdot}005 & 0{\cdot}005 \\ 0{\cdot}005 & 0{\cdot}005 \end{pmatrix}\begin{pmatrix} 2{\cdot}7 \\ 3{\cdot}4 \end{pmatrix}$$

$$\leqslant \begin{pmatrix} 0{\cdot}550 \\ 0{\cdot}460 \end{pmatrix} + \begin{pmatrix} 3{\cdot}355 \\ 2{\cdot}806 \end{pmatrix} = \begin{pmatrix} 3{\cdot}905 \\ 3{\cdot}266 \end{pmatrix}.$$

This result shows that since the coefficients of **A** and **b** are correct to two decimal places only, the solutions x_1 and x_2 are meaningless. To obtain meaningful solutions, correct to two decimal places say, the coefficients in the equations would have to be correct *to at least six decimal places*.

IV NUMERICAL SOLUTION OF DIFFERENTIAL EQUATIONS

An equation involving derivatives is called a differential equation.

For example $\dfrac{d^2y}{dt^2} + 2\dfrac{dy}{dt} + 3y = 0$ is a differential equation. The *order* of a differential equation is the order of the highest derivative of the unknown function in the equation. For example, the above equation is of order 2. The *solution* of the differential equation is an expression for the function $y(t)$ which is free of derivatives and which identically satisfies the equation.

There are differential equations which can be solved by analytical methods. For example

$$\frac{dy}{dt} - 2t = 0$$

has the solution

$$y(t) = \int 2t\,dt$$
$$= t^2 + c,$$

where c is a constant of integration. To determine c we need to know some initial condition, say $y(t) = y_0$ at $t = t_0$.

On the other hand, there are other differential equations, frequently occurring in practice, for which it is not possible to find simple analytical solutions, and a numerical method must be used. For example the equation

$$\frac{dy}{dt} = ty + t^2,$$

given that $y(t) = 0$ at $t = 0$,

98

has the solution

$$y(t) = e^{\frac{t^2}{2}} \int_0^t x^2 e^{-\frac{x^2}{2}} \, dx.$$

The integral on the right cannot be evaluated in terms of elementary functions, and so a numerical technique is required.

Numerical techniques

We consider the first order differential equation

$$y' = \frac{dy}{dx} = f(x, y) \qquad (4.1)$$

satisfying the initial condition

$$y = y_0 \quad \text{when} \quad x = x_0.$$

It will be assumed that f is a well-behaved function which in general implies that all the derivatives of f with respect to both x and y exist and are finite for all x and y in the domain of f.

By a numerical solution we mean approximating y at a number of tabular points, say at $x_0, x_1, \ldots, x_n$, where

$$x_r = x_0 + rh \quad (r = 0, 1, 2, \ldots, n)$$

and h is the constant spacing between the tabular points.

In this book we shall only consider the so-called *initial value* problems for which the initial conditions are all given at the one point x_0. This is in contrast to *boundary value* problems for which the initial conditions are given at more than just the one point. In this chapter we shall consider only a few of the many methods for solving differential equations.

4.1 Picard's method

Both Picard's and the Taylor series methods (the latter is discussed in Section 4.2) are useful for finding starting values, that is, values of the solution to the differential equation at a number of tabular points. These starting values are necessary when using some of the more sophisticated methods, for example the Adam–Bashforth method discussed in Section 4.4.

Although the Taylor series method is frequently used, sometimes, when for example the various derivatives are difficult to calculate or when the functions have a singularity about the point of interest, Picard's method may be very useful. The method is a simple iterative procedure explained and illustrated below.

If $y' = f(x, y)$ such that $y(x_0) = y_0$, then

$$y = y_0 + \int_{x_0}^{x} f(x, y)\, dx.$$

Picard's method consists of forming the sequence

$$y^{(n)} = y_0 + \int_{x_0}^{x} f(x, y^{(n-1)})\, dx \quad (n = 1, 2, \ldots).$$

Note

In the above formula $y^{(n)}$ are the successive Picard estimates of the solution and not the derivatives of y. This sequence of functions $y^{(0)}, y^{(1)}, y^{(2)}, \ldots$ converges to a limit function which is a particular solution of the differential equation if the following two conditions are satisfied:

(i) $f(x, y)$ is bounded in a suitable region about the point (x_0, y_0), i.e.

$$|f(x, y)| < K$$

for some positive number K in this region.

(ii) $$|f(x, y_1) - f(x, y_2)| < L|y_1 - y_2|$$

for some positive number L and any two points (x, y_1) and (x, y_2) having the same abscissas in the region considered.

Example 23

Find the first four Picard estimates (up to $y^{(4)}$) of the

100

solution to the differential equation

$$\frac{dy}{dx} = yx^{-\frac{1}{2}},$$

given that $y = 1$ at $x = 0$.

Solution

Note that since $x = 0$ is a singularity of $(yx^{-\frac{1}{2}})$, the Taylor series expansion about this point does not exist.

The first Picard estimate is given; it is

$$y^{(0)} = 1.$$

The subsequent estimates are:

$$y^{(1)} = 1 + \int_0^x y^{(0)}x^{-\frac{1}{2}}\,dx = 1 + \int_0^x 1.x^{-\frac{1}{2}}\,dx = 1 + 2x^{\frac{1}{2}},$$

$$y^{(2)} = 1 + \int_0^x (1 + 2x^{\frac{1}{2}})x^{-\frac{1}{2}}\,dx = 1 + 2x^{\frac{1}{2}} + 2x,$$

$$y^{(3)} = 1 + \int_0^x (1 + 2x^{\frac{1}{2}} + 2x)x^{-\frac{1}{2}}\,dx$$

$$= 1 + 2x^{\frac{1}{2}} + 2x + \tfrac{4}{3}x^{\frac{3}{2}},$$

$$y^{(4)} = 1 + \int_0^x (1 + 2x^{\frac{1}{2}} + 2x + \tfrac{4}{3}x^{\frac{3}{2}})\,dx$$

$$= 1 + 2x^{\frac{1}{2}} + 2x + \tfrac{4}{3}x^{\frac{3}{2}} + \tfrac{2}{3}x^2.$$

The pattern is clear; the nth estimate $y^{(n)}$ involves powers of x up to the order $\frac{n}{2}$. By choosing x small we can find an estimate of the solution to any specified accuracy.

4.2 A Taylor series method

If $y(x)$ is a well-behaved function whose successive derivatives at x_0 can be evaluated, we can approximate $y(x_0 + h)$ by a finite series

$$y(x_0 + h) \simeq y(x_0) + hy'(x_0) + \frac{h^2}{2!}y''(x_0) +$$

$$+ \frac{h^3}{3!}y'''(x_0) + \ldots + \frac{h^n}{n!}y^{(n)}(x_0). \quad (4.2)$$

The series on the right-hand side of equation (4.2) is known as a *truncated* Taylor series of *order n*. The series is of special importance in numerical analysis, not only as a practical method for solving differential equations but also as a basis for comparison with other methods for solving such equations. For example, we would say that some particular method is of order 2 if it approximates y_{n+1} by an equation equivalent to

$$y_{n+1} \simeq y_n + hy_n' + \frac{h^2}{2!}\, y_n''.$$

Note that the true value of y_{n+1}, given by the Taylor series is (using the order notation)

$$y_{n+1} = y_n + hy_n' + \frac{h^2}{2!}\, y_n'' + 0(h^3).$$

In general when the order of the method is n, the term $0(h^{n+1})$ is called the *truncation error* of the method.

In equation (4.2) $y(x_0) = y_0$ is the given initial condition. On setting $x = x_0$ and $y = y_0$ in equation (4.1) $y' = f(x, y)$ we can calculate $y'(x_0) = y_0'$. The values of the higher order derivatives $y_0^{(r)}$ $(r > 1)$ are found similarly by successively differentiating equation (4.1) and substituting in it the values obtained in the previous iterations. We denote the right-hand side of equation (4.2), which is the Taylor approximation to the true value of $y(x_0 + h) = y(x_1)$, by $Y(x_1) = Y_1$. Note that this notation does not indicate the order of the approximation but in general this is clear from the context. Generally we denote the Taylor approximation to $y(x_0 + rh) = y(x_r)$ by $Y(x_r) = Y_r$.

Having found the necessary derivatives at x_0, we can evaluate Y_1 using the nth order Taylor approximation, say. The next problem is to calculate Y_2. We can do this by using the same initial value as before, that is $y = y_0$ at $x = x_0$, and choose an interval $2h$ instead of h. But there are a number of disadvantages with this plan. First, to achieve the same accuracy as for Y_1, we would probably need to compute a

Taylor approximation of order greater than n. Secondly, the calculated coefficients (which would generally be stored) $\frac{h}{1}, \frac{h^2}{2!}, \frac{h^3}{3!}, \ldots$ would no longer be useful, since for an interval $2h$ the useful coefficients are $\frac{2h}{1}, \frac{(2h)^2}{2!}, \frac{(2h)^3}{3!}, \ldots$. We can avoid these disadvantages by choosing new 'initial conditions' by transferring the origin from x_0 to x_1. So to calculate Y_2 we use the right-hand side of equation (4.2), but for x_0 we use x_1 and for y_0 we use Y_1. This procedure is continued in an obvious way.

As an illustrative example we shall perform the first two steps in solving a differential equation which, for comparison, can be solved analytically in terms of elementary functions.

Example 24

Given

$$y'(x) = \frac{dy}{dx} = 2x + y$$

with initial values $x_0 = 0$, $y_0 = 1$, estimate $y = y(x)$ at $x = 0 \cdot 1$ and $x = 0 \cdot 2$ using the third order Taylor approximation.

Solution

The analytical solution to this equation is:

$$y = 3e^x - 2x - 2.$$

We are given

$$y'(x) = f(x, y) = 2x + y \quad \text{and} \quad h = 0 \cdot 1.$$

We calculate the various derivatives at x_0.

$$y'(x_0) = 2x_0 + y_0 = 0 + 1 = 1.$$

and on differentiation, we find

$$y''(x) = 2 + y'(x), \text{ so that } y''(x_0) = 2 + y'(x_0) = 3;$$
$$y'''(x) = y''(x), \qquad \text{so that } y'''(x_0) = y''(x_0) = 3,$$

and in general

$$y^{(r+1)}(x) = y^{(r)}(x) \text{ for all } r > 2.$$

The third order Taylor approximation then gives

$$y(0\cdot1) \simeq Y_1 = 1 + 0\cdot1 \times 1 + 0\cdot005 \times 3 + 0\cdot000\ 167 \times 3$$
$$= 1\cdot1155 \text{ (rounded to four decimal places).}$$

Now we change the origin so that effectively the new initial conditions are given by $x_0 = 0\cdot1, y_0 = 0\cdot1155$. We now find:

$$y'(x_0) = 0\cdot2 + 1\cdot1155 = 1\cdot3155,$$
$$y''(x_0) = 2 + 1\cdot3155 = 3\cdot3155,$$

and

$$y'''(x_0) = 3\cdot3155.$$

Hence

$$y(0\cdot2) \simeq Y_2 = 1\cdot1155 + 0\cdot1 \times 1\cdot3155 + 0\cdot005 \times 3\cdot3155 + {} + 0\cdot000\ 167 \times 3\cdot3155$$
$$= 1\cdot2642 \text{ (rounded to four decimal places).}$$

We can compare these approximations with exact solutions rounded to five decimal places. These are

$$y(0\cdot1) = 1\cdot115\ 51 \quad \text{and} \quad y(0\cdot2) = 1\cdot264\ 21.$$

It is seen that in this example the calculated results are very accurate. Nevertheless it is clear that if we continue to calculate the various Y_i, there will be a loss of accuracy as i increases.

For hand-calculating machines there is a simple running check which can be used during the calculations. If in the Taylor approximation for Y_{i+1} the sign of h is changed, the result is the approximation for $y(x_i - h) = y(x_{i-1}) = y_{i-1}$. Hence for a little extra computation, when calculating Y_{i+1} we can also find Y_{i-1} and so make sure that no arithmetical error has been made.

4.3 Euler's method

Euler's method is a special case of the Taylor series

approximation when only the first two terms of the series are used, i.e. it is a first order method.

If the initial conditions are given: $y = y_r$ at $x = x_r$, then the first order Taylor approximation for y_{r+1} is

$$Y_{r+1} = y_r + hy_r'$$
$$= y_r + hf(x_r, y_r). \qquad (4.3)$$

The last equation follows from equation (4.1) since $y_r' = f(x_r, y_r)$. Equation (4.3) has a simple geometrical interpretation. The graph shows the solution curve y and the given initial conditions (x_r, y_r).

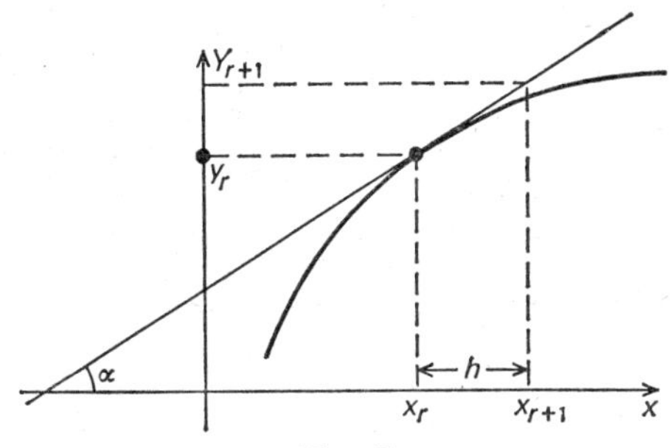

FIG. 7

In the interval (x_r, x_{r+1}) we approximate the solution curve by a straight line whose gradient is the same as that of the tangent at (x_r, y_r). If the gradient of this tangent is $\tan \alpha$, we have

$$\tan \alpha = f(x_r, y_r) \quad \text{(by equation 4.1)}.$$

Also, from the diagram

$$\tan \alpha = \frac{Y_{r+1} - y_r}{x_{r+1} - x_r} = \frac{Y_{r+1} - y_r}{h}$$

Hence $Y_{r+1} = y_r + hf(x_r, y_r)$, which is the equation (4.3).

The source of error in this method is clearly seen from the diagram. Not only is the approximation linear, but the gradient of the approximating line is that of the solution curve at *one end-point* of the interval. Generally this will not be the best or indeed a good linear approximation. Better results would be obtained if the gradient of the line were the average of the gradients of the solution curve at *both* endpoints of the interval. This is the basis of the so-called *modified Euler method.* To calculate this average gradient we need to know y_{r+1}, but this is what we are trying to evaluate. On the other hand we can approximate y_{r+1} by the Euler method, that is by

$$Y_{r+1} = y_r + hf(x_r, y_r).$$

We now have an approximation to the average gradient; it is

$$\tfrac{1}{2}[f(x_r, y_r) + f(x_{r+1}, Y_{r+1})].$$

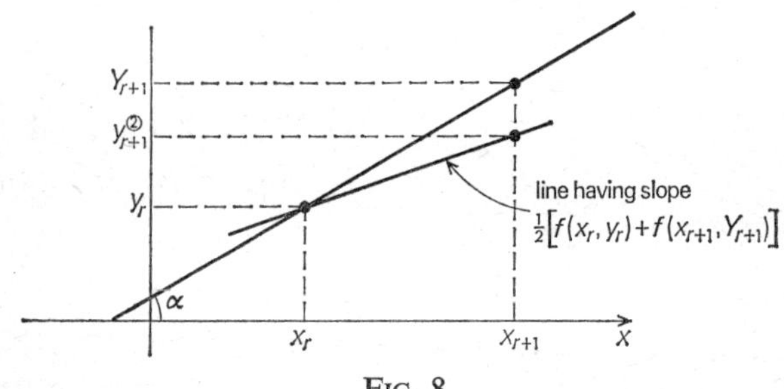

FIG. 8

We then obtain

$$y^{②}_{r+1} = y_r + \frac{h}{2}\,(f(x_r, y_r) + f(x_{r+1}, Y_{r+1})), \quad (4.4)$$

which in general is a better estimate of y_{r+1} than Y_{r+1}. It is not difficult to show that equation (4.4) is of order 2, so that the truncation error is smaller than the one for equation (4.3). Indeed, we can express the true values y_{r+1}, and $y'_{r+1} = f(x_{r+1}, y_{r+1})$ as Taylor series about x_r:

$$y_{r+1} = y_r + hy'_r + \frac{h^2}{2!}\,y''_r + \frac{h^3}{3!}\,y'''_r + \cdots,$$

$$y'_{r+1} = y'_r + hy''_r + \frac{h^2}{2!}\,y'''_r + \frac{h^3}{3!}\,y^{(iv)}_r + \cdots.$$

106

From the first equation, we obtain

$$y_{r+1} - y_r - hy'_r = \frac{h^2}{2} y''_r + \frac{h^3}{6} y'''_r + \dots,$$

and from the second

$$\frac{h}{2}(y'_{r \times 1} - y'_r) = \frac{h^2}{2} y''_r + \frac{h^3}{4} y'''_r + \frac{h^4}{12} y_r^{(iv)} + \dots.$$

On subtraction, the last two equations give us

$$(y_{r+1} - y_r) - \frac{h}{2}(y'_{r+1} + y'_r) = -\frac{h^3}{12} y'''_r - \dots$$

$$\simeq -\frac{h^3}{12} y'''_r \text{ (for small } h)$$

The above formula is equivalent to

$$y_{r+1} = y_r + \frac{h}{2} [f(x_r, y_r) + f(x_{r+1}, y_{r+1})] + 0(h^3),$$

hence it is an equation of order 2, the truncation error being of order 3.

The modified Euler is one of a group of techniques known as *predictor–corrector methods*. The basis of these methods is first to obtain an estimate of y_{r+1}, and then use this information to make a better estimate. The formula for the first estimate, in this case the formula for Y_{r+1}, is known as the *predictor*, and for the second estimate, for $y_{r+1}^{(2)}$, is known as the *corrector*.

Example 25

Solve for $x = 1 \cdot 06$ the differential equation

$$y' = xy^{\frac{1}{2}}$$

with initial condition: $y_0 = 1$ when $x_0 = 1$ using the modified Euler method.

Solution

We first use the predictor, equation (4.3), to find

$$Y_{r+1} = Y(1 \cdot 06) = 1 + (0 \cdot 06)(1) = 1 \cdot 06.$$

We next find

$$f(x_{r+1}, Y_{r+1}) = 1{\cdot}06(1{\cdot}06)^{\frac{1}{2}} = 1{\cdot}0918.$$

The modified Euler equation (4.4) is used as a corrector to yield

$$y^{(2)}(1{\cdot}06) = 1 + (0{\cdot}03)(1 + 1{\cdot}0918)$$
$$= 1{\cdot}0628 \text{ (to four decimal places).}$$

The process does not necessarily end at this stage. We use this last value of $y(1{\cdot}06)$ to find

$$f(1{\cdot}06, 1{\cdot}0628) = 1{\cdot}0929.$$

Reapplying the corrector we have

$$y(1{\cdot}06) = 1 + (0{\cdot}03)(1 + 1{\cdot}0929)$$
$$= 1{\cdot}0628 \text{ (to four decimal places).}$$

Since (to four decimal places) the last two iterations yield the same result, there is no point in continuing the process.

The above example illustrates how we use the predictor–corrector method to generate a sequence of numbers which converge to the solution (assuming of course that the process *does* converge).

4.4 The Adams–Bashforth method

We have established in Section 1.9 the following two formulae (equations (1.27) and (1.28)):

$$\int_{x_0}^{x_1} f \, dx = h[f_0 + \tfrac{1}{2}\nabla f_0 + \tfrac{5}{12}\nabla^2 f_0 + \tfrac{3}{8}\nabla^3 f_0 + \cdots]$$

and

$$\int_{x_0}^{x_1} f \, dx = h[f_1 - \tfrac{1}{2}\nabla f_1 - \tfrac{1}{12}\nabla^2 f_1 - \tfrac{1}{24}\nabla^3 f_1 - \cdots].$$

If $\dfrac{dy}{dx} = f(x, y)$, it follows that

$$\int_{x_0}^{x_1} f \, dx = y_1 - y_0,$$

so that the above two formulae can be written as

$$y_1 = y_0 + h[f_0 + \tfrac{1}{2}\nabla f_0 + \tfrac{5}{12}\nabla^2 f_0 + \tfrac{3}{8}\nabla^3 f_0 + \ldots] \qquad (4.5)$$

and

$$y_1 = y_0 + h[f_1 - \tfrac{1}{2}\nabla f_1 - \tfrac{1}{12}\nabla^2 f_1 - \tfrac{1}{24}\nabla^3 f_1 - \ldots] \qquad (4.6)$$

respectively.

The Adams–Bashforth method is one of the best known of the predictor–corrector methods, it uses equation (4.5) as the predictor and equation (4.6) as the corrector. One of the disadvantages of this and many other popular predictor–corrector methods is that they require a number of starting values to enable us to construct a difference table from which the appropriate backward differences can be found. The two equations (4.5) and (4.6) are used in a similar way to the predictor and corrector equations in example 25.

In general h is chosen small enough so as to make it unnecessary to use the corrector more than once—as in Example 25. The starting values are usually found either by the Taylor series approximation about the point x_0, or by Picard's method.

Example 26

Use the Adams–Bashforth method to solve the differential equation

$$\frac{dy}{dx} = 2x - y,$$

given that $y = 1$ at $x = 0$ for $x = 0.1(0.1)0.3$

Solution

By the Taylor series method (or Picard's method) we find

$$y = 1 - x + 3\frac{x^2}{2!} - 3\frac{x^3}{3!} + 3\frac{x^4}{4!}\ldots .$$

Since this example is used only as an illustration of the method, we shall use the above series to evaluate y at $x \pm 0.1$ and at $x \pm 0.2$ and construct a difference table with these values. We shall then use the Adams–Bashforth method to evaluate y at $x = 0.3$.

TABLE 19

	x	y	$\overset{.}{f}$	∇f	$\nabla^2 f$	$\nabla^3 f$
	-0.2	1.2642	-1.6642			
				3487		
	-0.1	1.1155	-1.3155		-332	
				3155		33
	0	1.0000	-1.0000		$-299.$	
				2856		26
	0.1	0.9144	-0.7144		-273	
				2583		31
	0.2	0.8561	-0.4561		-242	
				2341		25
(predicted)	0.3	0.8220	-0.2220		-248	
				2335		
(corrected)	0.3	0.8226	-0.2226			

Choosing $x_0 = 0.2$, we use the predictor formula, equation (4.5), to advance the solution to $x = 0.3$.

$$y_1 = y(0.3) = 0.8561 + 0.1[-0.4561 + \tfrac{1}{2} \times 0.2583 - \\ - \tfrac{5}{12} \times 0.0273 + \tfrac{3}{8} \times 0.0026]$$
$$= 0.8220.$$

We can now calculate the corresponding value of $f = 2x - y = 0.6 - 0.8220 = -0.2220$ and the various differences as shown in Table 19. We next use the corrector formula, equation (4.6) to calculate

$$y_1 = 0.8561 + 0.1[-0.2220 - \tfrac{1}{2} \times 0.2341 + \\ + \tfrac{1}{12} \times 0.0242 - \tfrac{1}{24} \times 0.0031]$$
$$= 0.8226.$$

We now have the corrected value of y_1, so we can evaluate the corresponding value of f and the differences as shown in Table 19. We use the corrector again, this time using the differences just calculated.

The new corrected value is $y = 0.8225$. This last value does not change in subsequent applications of the corrector.

In the above example it was necessary to use the corrector

twice (in fact three times, the third time was for checking purposes) to find the solution. As already mentioned, this is unusual and to avoid this we would have to choose a smaller h, say $h = 0.05$. We also could have obtained greater accuracy by using a larger number of terms in the predictor formula.

Because the Adams–Bashforth method requires at least two starting values of y, it is known as a *multi-step* method, this is in contrast to all the preceding methods discussed in this book which are *single-step* methods. Since the number of terms used in the predictor and the corrector formulae for the Adams–Bashforth method depends on whether the higher order differences used have any significant contribution to the result, we do not associate any truncation error with this method. On the other hand, since the coefficients in these formulae decrease slowly, there is a relatively large accumulation of rounding errors.

4.5 The Runge–Kutta method

The Runge–Kutta is a single-step method and hence does not require any special starting procedures. The calculations involved in using this method are fairly laborious, so the method is not recommended for desk computation; on the other hand it is well suited and widely used in automatic computation. Some of the advantages of this method are:

(i) it is not necessary to find the various repeated derivatives of the function.
(ii) the step length can be changed easily, and
(iii) the truncation error is very small (of order h^5).

In fact there is not just one but many so-called 'Runge–Kutta methods', that is, formulae for solving first order differential equations. But among these there is one used more frequently than all the others and so has become known as *the* Runge–Kutta method. Before discussing this popular fourth-order method we shall establish a simpler second-order equation, sometimes known as *Heun's method*. We assume that for the

111

differential equation

$$\frac{dy}{dx} = y' = f(x, y)$$

we have found (or are given)

$$y = y_r \quad \text{at} \quad x = x_r.$$

We wish to find a second order formula which will advance the solution to $x = x_{r+1} = x_r + h$. The second order Taylor series gives:

$$y_{r+1} = y_r + hy'_r + \frac{h^2}{2} y''_r + 0(h^3)$$

$$= y_r + hf(x_r, y_r) + \frac{h^2}{2} \frac{df}{dx}(x_r, y_r) + 0(h^3). \quad (4.7)$$

Since one of the objects of these methods is to avoid repeated differentiations, we must find an estimate of $\dfrac{df}{dx}$ in the above equation. For this purpose we consider a special case of the Taylor series for functions of two variables,

$$f(x + h, y + th) = f(x, y) + h \frac{\partial f}{\partial x}(x, y) +$$

$$+ th \frac{\partial f}{\partial y}(x, y) + 0(h^2). \quad (4.8)$$

At $x = x_r + h$ we estimate the value of y by the Euler equation, that is by

$$y_r + hf(x_r, y_r) = y_r + k_0 \text{ (say)}.$$

For this value of (x, y), equation (4.8) becomes:

$$f(x_r + h, y_r + k_0) = f(x_r, y_r) + h \frac{\partial f}{\partial x}(x_r, y_r) +$$

$$+ hf(x_r, y_r) \frac{\partial f}{\partial y}(x_r, y_r) + 0(h^2)$$

$$= f(x_r, y_r) + h \left(\frac{\partial f}{\partial x} + f \frac{\partial f}{\partial y} \right)(x_r, y_r) + 0(h^2)$$

$$= f(x_r, y_r) + h \frac{df}{dx}(x_r, y_r) + 0(h^2).$$

It follows from this equation that

$$\frac{df}{dx}(x_r, y_r) \simeq \frac{1}{h}\left[f(x_r + h, y_r + k_0) - f(x_r, y_r)\right].$$

Substituting this estimate in equation (4.7), we obtain

$$y_{r+1} = y_r + \frac{h}{2}\left[f(x_r, y_r) + f(x_r + h, y_r + k_0)\right] + 0(h^3).$$

Hence we have a second-order estimate for y_{r+1}. We write it in a simplified form as

$$Y_{r+1} = y_r + \tfrac{1}{2}(k_0 + k_1) \quad (r = 0, 1, 2, \ldots), \quad (4.9)$$

where
$$k_0 = hf(x_r, y_r)$$
and
$$k_1 = hf(x_r + h, y_r + k_0).$$

We can interpret geometrically the algorithm defined by equation (4.9) in the following way:

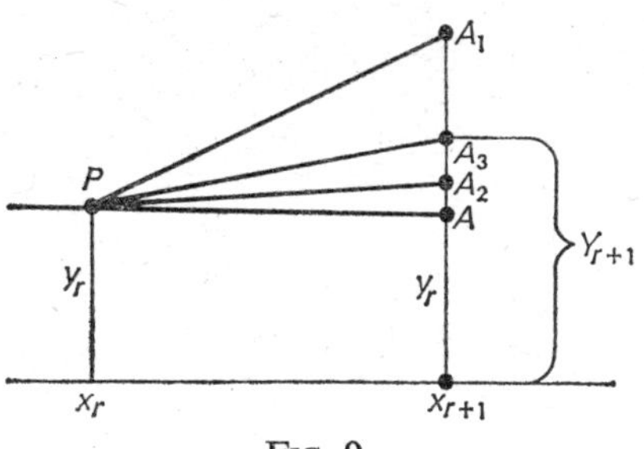

FIG. 9

From the point P, having known coordinates (x_r, y_r), we draw a line PA_1 with gradient $f(x_r, y_r)$. The increment AA_1 represents the length $hf(x_r, y_r) = k_0$.

Knowing this value of k_0, we can draw the line PA_2 with gradient $f(x_r + h, y_r + k_0)$. The increment AA_2 represents the length $hf(x_r + h, y_r + k_0) = k_1$. Now we can draw the line PA_3 such that the increment AA_3 in the average of the two increments AA_1 and AA_2, i.e. it represents the term $\tfrac{1}{2}(k_0 + k_1)$ in equation (4.9).

In a similar way to the above, although involving much

more manipulation, we can derive any of the Runge–Kutta formulae of higher order. In particular, the much used formula of order four is:

$$Y_{r+1} = y_r + \tfrac{1}{6}(k_0 + 2k_1 + 2k_2 + k_3) \quad (4.10)$$
$$(r = 1, 2, \ldots),$$

where
$$k_0 = hf(x_r, y_r),$$
$$k_1 = hf(x_r + \tfrac{1}{2}h, y_r + \tfrac{1}{2}k_0),$$
$$k_2 = hf(x_r + \tfrac{1}{2}h, y_r + \tfrac{1}{2}k_1),$$

and
$$k_3 = hf(x_r + h, y_r + k_2).$$

A derivation and a detailed discussion of the errors involved when using equation (4.10) can be found in reference [10]. Although equation (4.10) has the advantage that it can be evaluated relatively easily on a computer, it does involve evaluating f four times at each step. This can be a relatively lengthy operation compared, say, with a fourth-order predictor–corrector method which involves only two evaluations of f at each step. For this reason, and also because it is extremely difficult to estimate the error for a Runge–Kutta method (although it may actually be smaller than the corresponding error for the predictor–corrector method), it is advisable to use the Runge–Kutta method to start the solution and use a predictor–corrector method to continue it.

Example 27

Use the classical Runge–Kutta method to advance the solution of the differential equation

$$\frac{dy}{dx} = 2x + y, \text{ with } y(0{\cdot}1) = 1{\cdot}1155,$$

to $x = 0{\cdot}2$.

Solution

(Note that we have already considered this differential equation in Example 24.)

We have the following information:

$$f(x, y) = 2x + y, \ h = 0{\cdot}1, \ x_r = 0{\cdot}1, \ y_r = 1{\cdot}1155.$$

It follows that

$$k_0 = hf(x_r, y_r) = 0{\cdot}1(0{\cdot}2 + 1{\cdot}1155) = 0{\cdot}131\ 55,$$

$$k_1 = hf(x_r + \tfrac{1}{2}h, y_r + \tfrac{1}{2}k_0)$$
$$= 0{\cdot}1(0{\cdot}3 + 1{\cdot}181\ 27) = 0{\cdot}148\ 127,$$

$$k_2 = h(x_r + \tfrac{1}{2}h, y_r + \tfrac{1}{2}k_1)$$
$$= 0{\cdot}1(0{\cdot}3 + 1{\cdot}189\ 564) = 0{\cdot}148\ 956\ 4,$$

$$k_3 = h(x_r + h, y_r + k_2)$$
$$= 0{\cdot}1(0{\cdot}4 + 1{\cdot}264\ 456\ 4) = 0{\cdot}166\ 445\ 64.$$

Hence

$$y(0{\cdot}2) \simeq Y_{r+1} = 1{\cdot}1155 + \frac{0{\cdot}892\ 16}{6}$$
$$= 1{\cdot}2642 \text{ (to four decimal places).}$$

4.6 Differential equations of order higher than 1

We give a short account of one possible way of attempting to find a numerical solution of a differential equation of order higher than 1. A general nth-order differential equation can be written in the form:

$$\frac{d^n y}{dx^n} + a_{n-1}\frac{d^{n-1}y}{dx^{n-1}} + \ldots + a_1\frac{dy}{dx} + a_0 y = f(x, y). \quad (4.11)$$

(The a's may be functions of x and y.)

We can transform this nth-order equation into a system of n-simultaneous first-order differential equations by the following change of variables.

Let

$$y_1(x) = y,$$

$$y_2(x) = \frac{dy}{dx},$$

$$\vdots$$

$$y_n(x) = \frac{d^{n-1}y}{dx^{n-1}},$$

then equation (4.11) can be rewritten as the following system of equations:

$$\frac{dy_1}{dx} = y_2, \quad \frac{dy_2}{dx} = y_3, \quad \frac{dy_{n-1}}{dx} = y_n,$$

$$(4.12)$$

$$\frac{dy_n}{dx} = -a_{n-1}y_n - a_{n-2}y_{n-1} - \ldots - a_0 y_1 + f(x, y).$$

To find a solution to the differential equation (4.11) we must solve the system of equations (4.12). We can do this by one of the methods already discussed in this chapter. By 'advancing the solution of the system of equations (4.12) from x_0 to x_1' we mean 'evaluating the n-tuple of numbers $(y_1, y_2, \ldots, y_n)$ at x_1, given their value at x_0'.

Example 28 below indicates a method for doing this but does not go into much detail. Most books discussing the numerical solution of differential equations give more details, for example see reference [11].

Example 28

Write the differential equation

$$\frac{d^2 y}{dx^2} = a_0 + a_1 y + a_2 \frac{dy}{dx},$$

with initial conditions

$$y(x_0) = y_0 \quad \text{and} \quad y'(x_0) = y_0',$$

as a system of two first order differential equations and indicate how the solutions may be found at $x = x_0 + h$ and at $x = x_0 + 2h$.

Solution

Let $\qquad y_1 = y \quad \text{and} \quad y_2 = \frac{dy}{dx},$

then the differential equation becomes

$$\frac{dy_1}{dx} = y_2 \quad \text{(A) with initial conditions } y_1(x_0) = y_0,$$

116

$$\frac{dy_2}{dx} = a_0 + a_1 y_1 + a_2 y_2 \quad \text{(B)} \text{ with initial condition } y_2(x_0) = y_0'.$$

To find the solution we proceed as follows:

Step 1

To obtain $y_1(x_0 + h)$. We use equation (B) to find $y_2(x_0 + h)$ in terms of $y_1(x_0)$ and $y_2(x_0)$ (both given) by any numerical technique. We then use equation (A) to obtain $y_1(x_0 + h)$, again employing any numerical technique.

We now have both y_2 and y_1 at $x_0 + h$.

Step 2

Repeat the above procedure using equation (B) to obtain $y_2(x_0 + 2h)$ in terms of $y_2(x_0 + h)$ and $y_1(x_0 + h)$ which were determined in step 1. Use equation (A) to calculate $y_1(x_0 + 2h)$.

4.7 Accuracy and stability

We have already considered some aspects of the accuracy of a number of methods for solving differential equations discussed in this book. It must be stated that certain aspects of accuracy and error analysis in this field are extremely complicated and indeed are the subjects of much present day research in Numerical Analysis. The type of errors to be considered are blunders, round-off errors, truncation errors and others.

Now the fact that, in general, errors of some type are bound to arise in the numerical solution of differential equations is accepted as unavoidable. But for these errors to grow rapidly, say exponentially, and so possibly cause what is known as *instability* is unacceptable. But this is exactly what may happen when using some of the methods discussed in this book and indeed with many other methods not discussed here. To analyse the instabilities which may arise in practice is beyond the scope of this book, but at least one type of instability will be considered briefly. This instability arises because the equation we are solving numerically is not a differential but a *difference* (or *recurrence*) *equation.* For example, consider the numerical solution of the differential equation

$$\frac{dy}{dx} = f(x, y) \qquad (4.13)$$

by the Taylor approximation of the second order

$$y_{r+1} = y_r + hy'_r + \frac{h^2}{2} y''_r$$

(for this analysis we do not distinguish between y_{r+1} and Y_{r+1}). We have

$$y'_r = f(x_r, y_r) = f_r, \quad \text{and} \quad y''_r = f'_r,$$

so that we can write the Taylor approximation in the form

$$y_{r+1} = y_r + \left(hf_r + \frac{h^2}{2} f'_r \right), \qquad (4.14)$$

which is a recurrence relation of the first order. So in fact the equation we are solving is not equation (4.13) but an approximation, equation (4.14).

Similarly, some of the numerical methods for solving differential equations lead to considerations of difference equations of order 2, or even of higher orders. A first-order difference equation has the general form

$$ay_{r+1} + by_r = c_r,$$

and a difference equation of order 2 has the form

$$ay_{r+2} + by_{r+1} + cy_r = d_r. \qquad (4.15)$$

The general solution of equation (4.15) consists of two parts; one is any particular solution and the second is the complementary solution. This is analogous to the solution of a differential equation.

If $y_r^{(0)}$ is any particular solution to equation (4.15) and α, β are the roots of the auxiliary equation

$$am^2 + bm + c = 0 \qquad (4.16)$$

then the general solution to equation (4.15) has the form

$$y_r = A_1\alpha^r + A_2\beta^r + y_r^{(0)}, \qquad (4.17)$$

where A_1 and A_2 are arbitrary constants determined by the initial conditions. It is intuitively clear that the complementary solution of the difference equation plays a crucial role in considering the stability of the method being examined. Let us take our intuitive argument a few steps further. If the differential equation is of order 1, and the method we are using leads to a difference equation of the first order, then both types of equations have one complementary solution. We would hope that the two complementary solutions would be of similar type, in which case the solution of the difference equation would hopefully be a good approximation to the solution of the differential equation, and the method is stable.

Suppose next that the method we are using for solving a first order differential equation leads to a difference equation of order 2. We now have *one* complementary solution for the differential equation and *two* complementary solutions, their values depending on the solution of equation (4.16), for the difference equation. Let us agree for the sake of argument that one of the complementary solutions of the difference equation corresponds closely to the complementary solution of the differential equation. We then have to investigate the effect on equation (4.17) of the second complementary solution, known as the *parasitic solution*. It may be that as r increases the contribution due to the parasitic solution decreases or remains constant or indeed increases very slowly. In these cases we would consider the method under investigation to be *stable*. On the other hand the contribution of the parasitic solution may be such that it swamps the contribution of the required solution as r increases. In this case the method under investigation is *unstable*.

It should be appreciated that even if both the differential and the difference equations are of the first order, the method being investigated may be unstable, since the

complementary solutions of the difference equation may be
of a different type to that of the differential equation.

It is hoped that the above intuitive arguments have given
some indication of what is meant by stability and non-
stability of a method for solving differential equation. An
interesting and rigorous discussion of this topic can be
found in reference [4].

V   NON-LINEAR EQUATIONS

In Chapter III we discussed linear equations. Any equation or function which is not of that type, we call *non-linear*.

In this chapter we shall consider the problem of finding the number x_0 (or the set of numbers $\{x_0\}$) for which

$$f(x_0) = 0$$

for some non-linear function f. This number x_0 (or the set of numbers) is called a *zero* (or a *root*) of f.

The solutions to many problems in engineering and more generally in mathematics, with particular reference to Numerical Analysis itself, ultimately depend on finding the roots of some function: in particular on finding the roots of a polynomial function.

We have discussed in the previous chapters both direct and iterative numerical methods. All the numerical methods for finding zeros of functions are iterative procedures. Indeed it is well known that there exist closed formulae, that is direct methods, for finding roots of polynomial functions of degree four at most. The formulae for the roots of cubic and quartic polynomials are rather complicated and are very rarely used in practice, so generally all numerical methods for finding the roots of polynomials of degree greater than two are indirect.

One very useful 'non-numerical' and direct method for finding the roots is the graphical method. Unfortunately it is not always easy to sketch the graph of a function, and even when it is done, the accuracy is usually rather poor. Nevertheless it is very useful to find the approximate roots of a polynomial by a graphical method and then to improve the approximation by an iterative procedure.

5.1 Linear iterations

The problem is to find the root of

$$f(x) = 0. \tag{5.1}$$

We rewrite equation (5.1) in the form

$$x = F(x), \tag{5.2}$$

so that any $x = a$ satisfying equation (5.2) is also a solution to equation (5.1). The rearrangement of equation (5.1) into the form of equation (5.2) is not unique and some rearrangements for a particular problem result in better iterative procedures than others. Indeed for a particular linear iteration method, one rearrangement may lead to a converging sequence, while another one may not.

Example 29

Find three different rearrangements of the equation

$$x^3 - 3x - 1 = 0$$

in the form of equation (5.2).

Solution

The cubic equation above is considered in some detail in Section 2.4 of reference [3]. One possible rearrangement of the equation is

$$x = \frac{x^3 - 1}{3}.$$

Two other rearrangements, obtained by dividing f by x and x^2 respectively (assuming that $x \neq 0$), are:

$$x = \frac{1}{x^2 - 3}$$

and

$$x = \frac{3x + 1}{x^2}.$$

Equation (5.2) has a simple geometrical interpretation. The

122

solution, that is the root of the equation (5.1), is the value of x at which the line $y = x$ intersects the curve $y = F(x)$: There may be no point or one point or more than one point of intersection. In the first case, when there is no point of intersection, the roots of equation (5.1) are complex. We shall assume that there exists at least one point of intersection, say at $x = \alpha$ and that in the neighbourhood of this point both $F(x)$ and $F'(x)$ are continuous.

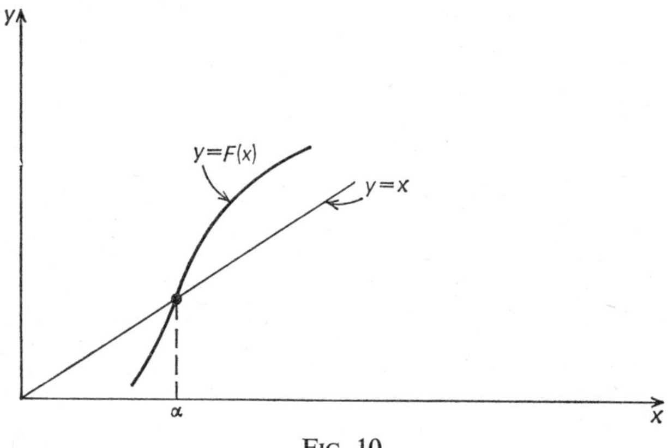

Fig. 10

The iteration we use is a very simple one: it is

$$x_{i+1} = F(x_i) \quad (i = 0, 1, 2, \ldots). \tag{5.3}$$

To start off the iterations defined by equation (5.3) we choose x_0 in the neighbourhood of $x = \alpha$. This approximation to α can be found by a graphical method (say).

If the iterations converge to α then

$$\alpha = \lim_{n \to \infty} x_{n+1} = \lim_{n \to \infty} F(x_n) = F(\lim_{n \to \infty} x_n) = F(\alpha).$$

It follows that $x = \alpha$ satisfies equation (5.2) and hence is a root of equation (5.1).

In Example 30 below it is shown that the iterations defined by equation (5.3) do not always converge.

Example 30

The roots of the equation (considered in Example 29)

$$f(x) = x^3 - 3x - 1 = 0$$

are approximately $-1 \cdot 5$, $-0 \cdot 3$ and $1 \cdot 9$. Use these approximations as starting values in the rearrangement of the equation in the form

$$x = \frac{x^3 - 1}{3}$$

to find three sequences. Discuss the results.

Solution

This problem is considered in some detail in reference [3]. It is shown that, starting with the three approximations, the following sequences are obtained:

(1) Starting with $x_0 = -1 \cdot 5000$

$$-1 \cdot 5000$$
$$-1 \cdot 4583$$
$$-1 \cdot 3672$$
$$-1 \cdot 1851.$$

(2) Starting with $x_0 = -0 \cdot 3000$

$$-0 \cdot 3000$$
$$-0 \cdot 3423$$
$$-0 \cdot 3467$$
$$-0 \cdot 3472$$
$$-0 \cdot 3473$$
$$-0 \cdot 3473.$$

(3) Starting with $x_0 = 1 \cdot 9000\ 00$

$$1 \cdot 9000$$
$$1 \cdot 9530$$
$$2 \cdot 1497$$
$$2 \cdot 4714$$

If the three roots of f are α_1, α_2, α_3 close to $-1\cdot5$, $-0\cdot3$ and $1\cdot9$ respectively, the above example shows that the rearrangement results in only one of the sequences, the second one, converging to the root nearest to the starting value, in this case $\alpha_2 = -0\cdot3473$. In fact it will be found that the first sequence also converges to this root, but in this instance we would say that the rearrangement has failed since the root we are after is α_1, which is near $-1\cdot5$.

It is of obvious interest to find out why the iterations are sometimes successful and sometimes fail. The explanation is given in reference [3] in terms of interval magnification. We now use the mean value theorem of the differential calculus to suggest an explanation in slightly different terms.

We assume that α is a root of equation (5.1) so that

$$\alpha = F(\alpha).$$

On subtracting this equation from equations (5.3) and using the mean value theorem we obtain

$$\begin{aligned}
x_{i+1} - \alpha &= F(x_i) - F(\alpha) \\
&= (x_i - \alpha)F'(\eta_i) \text{ for some } \eta_i \in [x_i, \alpha] \\
&\quad (i = 0, 1, 2, \ldots).
\end{aligned} \tag{5.4}$$

If $F'(\eta_i)$ are (1) continuous and (2) smaller than unity in modulus in the interval involved, then it follows from equation (5.4) that these intervals are such that

$$|x_0 - \alpha| > |x_1 - \alpha| > |x_2 - \alpha| > \ldots,$$

and it is clear that the sequence $\{x_i\}$ converges to α.

The last statement can be proved rigorously in the following way. We define the *error* in the *i*th iteration as $r_i = x_i - \alpha$. With this notation equation (5.4) becomes

$$r_{i+1} = F'(\eta_i)r_i \quad (i = 0, 1, 2, \ldots).$$

If for all i, $F'(\eta_i) \leqslant L < 1$, then

$$|r_{i+1}| \leqslant L|r_i| \leqslant L^2|r_{i-1}| \leqslant \ldots \leqslant L^{i+1}|r_0|.$$

Since $|L| < 1$, it follows that

$$\lim_{i \to \infty} |r_{i+1}| = \lim_{i \to \infty} L^{i+1}|r_0| = 0,$$

hence

$$\lim_{i \to \infty} x_i = \alpha.$$

The above analysis demonstrates that so long as $F'(x)$ is continuous and in modulus smaller than unity at each point in an interval containing the root α and the initial guess x_0, the sequence $\{x_i\}$ converges to α.

By a similar argument it can be shown that if $|F'(x)|$ is greater than unity in an interval about the root α, the sequence $\{x_i\}$ does not converge to α, although it may converge to another root of f.

Example 31

Apply the above analysis to the rearrangement of f suggested in example 30.

Solution

Since

$$F(x) = \frac{x^3 - 1}{3}, \quad F'(x) = x^2.$$

At both $x = -1\!\cdot\!5$ and $x = 1\!\cdot\!9$,

$$|F'(x)| > 1,$$

hence we would not expect the sequences $\{x_i\}$ to be convergent to the respective roots α_1 and α_3 in either case.

On the other hand, when $x = -0\!\cdot\!3$,

$$|F'(x)| < 1,$$

and we would expect the sequence to converge to the root α_2 near $x = -0\!\cdot\!3$.

The above discussion has a simple geometrical interpretation. We consider two different functions $F(x)$ whose graphs are shown below.

The point of intersection of the line $y = x$ with the curve

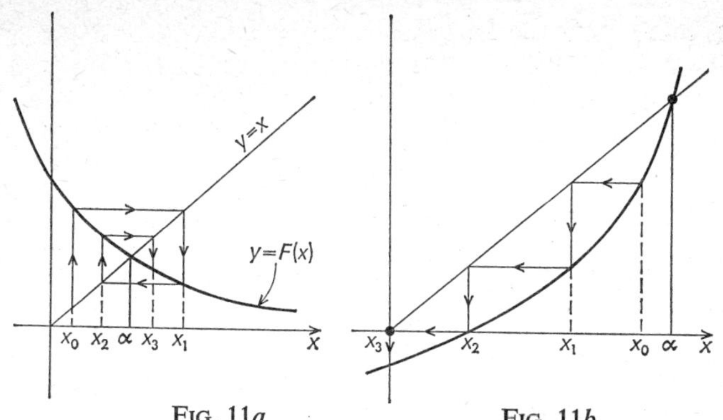

FIG. 11*a*   FIG. 11*b*

$y = F(x)$ is marked in both diagrams as having abscissa α. In figure 11*a* it is clear that $|F'(x)| < 1$ in the neighbourhood of $x = \alpha$. Our linear iterative algorithm produces the sequence of points $x_0, x_1, x_2, \ldots$, which in figure 11*a* is seen to converge to α whereas in figure 11*b* the sequence is seen to diverge from α.

Speed of Convergence

We shall now discuss another important property of algorithms for finding roots of an equation. Assuming that the sequence of iterations does converge to the root α, how fast is the convergence? It is convenient for this investigation to introduce the concept of the *order* of an algorithm.

We consider the errors r_i and r_{i+1} (defined above) in two successive iterations. If for a particular algorithm

$$\lim_{i \to \infty} \frac{|r_{i+1}|}{|r_i|^n} = C \qquad (n = 1, 2, \ldots), \qquad (5.5)$$

where C is a constant depending on f, then we say that the algorithm is of order n. We shall only be concerned in this book with linear and quadratic algorithms, that is, with algorithms of order 1 and 2.

It is important to understand the implications of equation (5.5). For large i, the equation can be written as

$$|r_{i+1}| \simeq C|r_i|^n.$$

If the error in the ith iteration r_i is small, then the error in the $(i + 1)$th iteration r_{i+1} is extremely small, being proportional to r_i *raised to the power n* (n being assumed greater than unity, say).

In the case of the iterations considered in this section we have from equation (5.4)

$$r_{i+1} = F'(\eta_i)r_i,$$

hence

$$\lim_{i \to \infty} \frac{|r_{i+1}|}{|r_i|} = \lim_{i \to \infty} |F'(\eta_i)| = |F'(\alpha)|$$

(since $F'(x)$ is assumed continuous)

$$= C.$$

This implies that $n = 1$, and hence the iterations are of order 1, i.e. are *linear*. The last equation also shows that the greater the value of the constant C the slower the convergence of the iterations, and if C is greater than unity there will be no convergence. There exist a number of methods for speeding up the convergence of the iteration; for further information about them see (for example) reference [10].

5.2 Quadratic iterations

We shall again assume that the functions we are dealing with are 'well-behaved'. Among the best-known algorithms of the second order for finding the roots of an equation

$$f(x) = 0 \tag{5.6}$$

is the *Newton–Raphson* method.

Let x_0 be an approximation to the root α of the equation (5.6), and let h_1 be the corresponding error, so that

$$\alpha = x_0 + h_1.$$

Using the Taylor series we can write:

$$0 = f(\alpha) = f(x_0 + h_1) = f(x_0) + h_1 f'(x_0) + \tfrac{1}{2}h_1^2 f''(x_0) + \cdots.$$

For small h_1, the equation becomes

$$0 \simeq f(x_0) + h_1 f'(x_0),$$

so that $h_1 \simeq -\dfrac{f(x_0)}{f'(x_0)}$, so long as $f'(x_0) \neq 0$.

It follows that a better approximation for a is

$$x_1 = x_0 - \frac{f(x_0)}{f'(x_0)}.$$

On repeating the process we obtain:

$$x_{i+1} = x_i - \frac{f(x_i)}{f'(x_i)} \qquad (i = 0, 1, 2, \ldots). \qquad (5.7)$$

Example 32

Find the square root of 5.

Solution

It is instructive to consider how to use the algorithm defined by equation (5.7) to find the square root of a positive number N. The square root of N is the positive root of the equation

$$f(x) = x^2 - N.$$

Since in this case $f'(x) = 2x$, the algorithm, equation (5.7), becomes

$$x_{i+1} = x_i - \frac{x_i^2 - N}{2x_i}$$

$$= \tfrac{1}{2}\left[x_i + \frac{N}{x_i}\right] \qquad (i = 0, 1, 2, \ldots).$$

Starting with $x_0 = 2$, we obtain the following successive approximations:

$$2 \cdot 000$$
$$2 \cdot 250$$
$$2 \cdot 235$$
$$2 \cdot 236.$$

In the above example the convergence is seen to be very fast. To compare this speed of convergence with the speed of convergence of the method considered in the previous section, we need to determine the order of Newton's algorithm.

For the purpose of this analysis we assume that $f'(\alpha) \neq 0$. As defined previously, let the error in the ith iteration be

$$r_i = x_i - \alpha,$$

so that

$$r_{i+1} = x_{i+1} - \alpha = x_i - \frac{f(x_i)}{f'(x_i)} - \alpha$$

$$= r_i - \frac{f(x_i)}{f'(x_i)}. \qquad (5.8)$$

As $x_i = r_i + \alpha$, we can use the Taylor series to find

$$\frac{f(x_i)}{f'(x_i)} = \frac{f(\alpha + r_i)}{f'(\alpha + r_i)} = \frac{f(\alpha) + r_i f'(\alpha) + \frac{1}{2} r_i^2 f''(\alpha) + \ldots}{f'(\alpha) + r_i f''(\alpha) + \frac{1}{2} r_i^2 f'''(\alpha) + \ldots}.$$

Since $f(\alpha) = 0$, equation (5.8) can be written as

$$r_{i+1} = r_i - \frac{r_i f'(\alpha) + \frac{1}{2} r_i^2 f''(\alpha) + \ldots}{f'(\alpha) + r_i f''(\alpha) + \frac{1}{2} r_i^2 f'''(\alpha) + \ldots}$$

$$= \frac{r_i [f'(\alpha) + r_i f''(\alpha) + \ldots] - [r_i f'(\alpha) + \frac{1}{2} r_i^2 f''(\alpha) - \ldots]}{f'(\alpha) + r_i f''(\alpha) + \frac{1}{2} r_i^2 f'''(\alpha) + \ldots}$$

$$\simeq \frac{\frac{1}{2} r_i^2 f''(\alpha)}{f'(\alpha)},$$

where we have neglected all terms involving higher powers of r_i in the numerator and all terms involving factors of r_i in the denominator (with respect to the term $f'(\alpha)$). From the previous equation we find that

$$\lim_{i \to \infty} \frac{|r_{i+1}|}{|r_i|^2} = \lim_{i \to \infty} \frac{1}{2} \frac{f''(\alpha)}{f'(\alpha)} = C \quad \text{(a constant)}.$$

This implies that the algorithm is of order 2, i.e. *quadratic*. If in the above equation we assume that $C \simeq 1$, then for large i

$$r_{i+1} = r_i^2 \quad \text{(approximately)}.$$

This shows that for a second-order convergence method the

130

number of correct decimal places is approximately doubled at each iteration. For example if

$$|r_i| = |x_i - \alpha| = \tfrac{1}{10},$$

then

$$|r_{i+1}| = |x_i - \alpha|^2 = \tfrac{1}{100}.$$

For further information and greater detail see reference [10].

5.3 Bairstow's method

The Newton–Raphson method can be modified to evaluate complex roots if complex arithmetic is used and the initial approximation is a complex number. To find the complex (and the real) roots of a polynomial equation having real coefficients without using the rather cumbersome and time-consuming complex arithmetic, use is made of Bairstow's method, which will now be described. If the polynomial of degree n in question is

$$f(x) = a_n x^n + a_{n-1} x^{n-1} + a_{n-2} x^{n-2} + \ldots + a_0,$$

having the complex root $\alpha = a + ib$, then the complex conjugate $\bar{\alpha} = a - ib$ is also a root, and the real quadratic polynomial $x^2 + sx + t = (x - \alpha)(x - \bar{\alpha})$ is a factor of $f(x)$.

We choose a first approximation to the coefficients of the quadratic polynomial and by an iterative procedure improve the approximations until a specified accuracy is achieved.

Let p and q be the first approximations to the coefficients of the quadratic polynomial. We can write

$$f(x) = (x^2 + px + q)g(x) + r_1 x + r_0, \qquad (5.9)$$

where $g(x)$ is the quotient polynomial. It is of degree $(n - 2)$, say

$$g(x) = b_n x^{n-2} + b_{n-1} x^{n-3} + \ldots + b_2,$$

and $(r_1 x + r_0)$ is the remainder after $f(x)$ is divided by $(x^2 + px + q)$.

On equating the coefficients of equations (5.9) we find

$$b_s = a_s - pb_{s+1} - qb_{s+2} \qquad (5.10)$$
$$(s = n, n-1, \ldots, 2),$$

where we make

$$b_{n+1} = b_{n+2} = 0.$$

Using the recurrence equation (5.10) we define two coefficients b_1 and b_0 as

$$b_1 = a_1 - pb_2 - qb_3$$

and
$$b_0 = a_0 - pb_1 - qb_2. \qquad (5.11)$$

Also, on equating the coefficients of x^1 and x^0 in equation (5.9), we find

$$a_1 = pb_2 + qb_3 + r_1$$

and
$$a_0 = qp_2 + r_0. \qquad (5.12)$$

By simple manipulation of equations (5.11) and (5.12) we find

$$b_1 = r_1$$

and
$$b_0 + pb_1 = r_0. \qquad (5.13)$$

We next divide the polynomial $g(x)$ by $(x^2 + px + q)$ and obtain a quotient polynomial $h(x)$ of degree $(n-4)$ and a remainder $S_1 x + S_0$ so that

$$g(x) = (x^2 + px + q)h(x) + S_1 x + S_0, \qquad (5.14)$$

where

$$h(x) = C_{n-2} x^{n-4} + C_{n-3} x^{n-5} + \ldots + C_2. \qquad (5.15)$$

The equivalent recurrence relation to the equation (5.10) is

$$C_s = b_{s+2} - pC_{s+1} - qC_{s+2} \qquad (5.16)$$
$$(s = n, n-1, \ldots, 2),$$

where

$$C_{n+1} = C_{n+2} = 0,$$

and the equivalent relation to the equations (5.13) are
$$C_1 = S_1$$
and
$$C_0 + pC_1 = S_0. \tag{5.17}$$
On combining equations (5.9) and (5.14) we obtain
$$f(x) = (x^2 + px + q)^2 h(x) + \\ + (x^2 + px + q)(S_1 x + S_0) + r_1 x + r_0. \tag{5.18}$$
Next we assume that increments Δp in p and Δq in q (note here that the Δ notation is used as an increment and not as a finite difference) makes the quadratic polynomial a factor of $f(x)$, i.e.
$$x^2 + (p + \Delta p)x + q + \Delta q \text{ is a factor of } f(x).$$
If α (assumed complex) is a root of the above factor, it is also a root of f. We then have
$$\alpha^2 + (p + \Delta p)\alpha + q + \Delta q = 0,$$
and hence
$$(\alpha^2 + p\alpha + q) = -(\Delta p \alpha + \Delta q). \tag{5.19}$$
It follows that $(\alpha^2 + p\alpha + q)^2$ is of second order in Δp and Δq and hence can be neglected in the expression for $f(a)$ obtained by making $x = \alpha$ is equation (5.18).

Neglecting terms of second order in Δp and Δq and making use of equation (5.19), we obtain the following approximate expression from equation (5.18):
$$(S_1 p \Delta p - S_1 \Delta q - S_0 \Delta p + r_1)\alpha + \\ + (q S_1 \Delta p - S_0 \Delta q + r_0) = 0.$$

This last (approximate) equation must in general hold for two values of α (the quadratic factor of $f(x)$ involves two roots of that polynomial), so that we must have
$$(pS_1 - S_0)\Delta p - S_1 \Delta q + r_1 = 0$$
and
$$q S_1 \Delta p - S_0 \Delta q + r_0 = 0.$$

Solving these two simultaneous equations for Δp and Δq, we find
$$\Delta p = \frac{r_1 S_0 - r_0 S_1}{d}$$

and
$$\Delta q = \frac{r_0 S_0 + q r_1 S_1 - p r_0 S_1}{d},$$
(5.20)

where
$$d = S_0{}^2 - p S_1 S_0 + q S_1{}^2.$$

From equations (5.20) we can calculate the increments which will give a closer approximation to a quadratic factor of the polynomial $f(x)$.

At each iteration new values of p and q are found, and using equations (5.13) and (5.17) new values of r_0, r_1, S_0 and S_1 are calculated. The iterations are continued until Δp and Δq are found to be zero (within specified tolerances).

Example 33

$(x^2 + x + 1)$ is an approximate factor of the polynomial equation
$$f(x) = x^4 - 4x^3 + 9x^2 - 4x + 8.$$

Use Bairstow's method to improve this approximation.

Solution

Since $f(x)$ is a polynomial of degree 4, equations (5.9) must be of the form
$$f(x) = (x^2 + x + 1)(b_4 x^2 + b_3 x + b_2) + r_1 x + r_0.$$

The coefficients of $f(x)$ are:
$$a_4 = 1$$
$$a_3 = -4$$
$$a_2 = 9$$
$$a_1 = -4$$
$$a_0 = 8,$$

so that we can calculate the coefficients of g with the aid of equation (5.10):
$$b_4 = 1$$
$$b_3 = -5$$
$$b_2 = 13,$$

also
$$b_1 = -12$$
and
$$b_0 = 7.$$

134

It follows (equation 5.13) that

$$r_1 = -12 \quad \text{and} \quad r_0 = -5.$$

We can write the polynomial g in the form of equation (5.14) but for this simple example we can find all the coefficients by inspection. Indeed

$$g(x) = x^2 - 5x + 13 = (x^2 + x + 1)(1) - 6x + 12,$$

so that

$$S_1 = -6 \quad \text{and} \quad S_0 = 12.$$

We can now evaluate Δp and Δq by use of equations (5.20)

$$\Delta p = -\tfrac{174}{252} \simeq -0{\cdot}69$$

and

$$\Delta q = -\tfrac{18}{252} \simeq -0{\cdot}07,$$

so that the next approximation to the quadratic factor is

$$(x^2 + 0{\cdot}31x + 0{\cdot}93).$$

In fact the polynomial $f(x)$ can be written in terms of its quadratic factors as $f(x) = (x^2 + 1)(x^2 - 4x + 8)$. It is seen that the first iteration alone brings us much nearer the desired factor. In fact the iterative process is, like the Newton–Raphson's method, quadratically convergent (see reference [1]).

Having computed the quadratic factor of f to a specified accuracy, we determine the corresponding two roots by direct calculation. The whole process is then repeated with the remainder polynomial of degree $(n - 2)$.

CONCLUSION

When writing a relatively short book such as this, it is very difficult both to select the techniques to be discussed and to judge the right balance between the intuitive and the rigorous approach.

The selection of the material, especially the emphasis on Finite Differences in Chapter I, was based on academic rather than practical reasoning.

Although a fairly rigorous error analysis has been carried out for some of the techniques discussed, frequently reference has been made to more advanced texts which do this particular job well. But it is sincerely hoped that, having read the book, the reader appreciates the importance of error analysis in Numerical Analysis.

Finally, the author will have achieved his objectives in writing this book if, after having read it, the reader's interest in Numerical Analysis is increased to the extent that he will wish to study more advanced works on this subject, some of which are mentioned in the Bibliography.

BIBLIOGRAPHY

[1] *Modern Computing Methods* (Her Majesty's Stationery Office).
[2] *Interpolation and Allied Tables* (Her Majesty's Stationery Office).
[3] A. Graham and G. A. Read, *Calculus via Numerical Analysis*, Transworld Student Library, 1972.
[4] L. Fox and D. F. Mayers, *Computing Methods for Scientists and Engineers*, O.U.P., 1968.
[5] L. Fox, *An Introduction to Numerical Linear Algebra*, O.U.P., 1964.
[6] P. J. Davis, *Interpolation and Approximation*, Blaisdell Publishing Co., 1963.
[7] D. K. Faddeev and V. N. Faddeeva, *Computational Methods of Linear Algebra*, W. H. Freeman and Co., 1963.
[8] C. E. Froberg, *Introduction to Numerical Analysis*, Addison-Wesley Publishing Co., 1970.
[9] F. B. Hildebrand, *Principles of Numerical Analysis*, McGraw-Hill Book Co., 1953.
[10] A. Ralston, *A First Course in Numerical Analysis*, McGraw-Hill Book Co., 1965.
[11] J. Singer, *Elements of Numerical Analysis*, Academic Press, 1964.

Information about other books
in the Transworld Student
Library will be found on the
following pages.

Transworld Student Library

General Editor:
H. GRAHAM FLEGG, M.A., D.C.Ae., C.Eng., F.I.M.A.,
M.I.E.E., A.F.R.Ae.S., F.R.Met.S.
Reader in Mathematics, The Open University

Transworld Student Library is a series of paperback books devoted to topics in Mathematics and the Sciences designed to meet the modern educational needs of the independent learner. The academic level varies with the familiarity of the material covered, from general introductory presentations suitable for sixth-form pupils and school leavers to more specialized topics treated at first- or second-year University level. At the same time, many of the books will be of interest to the general reader with an enquiring mind who wishes to become acquainted with some of the more recent developments in Mathematics and the Sciences.

For the most part, the treatment breaks away from the traditional presentation geared to the classroom situation and provides a refreshingly new approach to the subject matter being discussed.

The criteria for inclusion in the series are that a book shall be clear and understandable in what it has to say; that the academic standard shall be impeccable; that the author shall be genuinely enthusiastic in his desire and ability to communicate to the reader; that the presentation shall be as concise as is compatible with clear understanding; that particular attention shall be paid to the provision of illustrative material; and that the principal aim shall be to stimulate the reader's interest and his desire to study further as well as to provide information and general background material.

Some of the books in the Library are first English translations of outstanding foreign works where these meet the criteria for inclusion, but the majority are specially commissioned from authors who, whilst being specialists in their subjects, are nevertheless prepared to break away from traditional and now outmoded approaches and present their material in a manner consistent with the new adult educational requirements.

THEORETICAL STATISTICS—BASIC IDEAS   70p
0 552 40002 5   100 pages

Stanley N. Collings
Reader in Statistics, The Open University

Theoretical Statistics does not pretend that statistics is a non-mathematical subject. Starting from a few A-level concepts, and confining itself to discrete situations, it provides a clear introduction to how these concepts are used in formulating the basic ideas upon which probability and sampling notions are built.

BOOLEAN ALGEBRA   80p
0 552 40001 7   160 pages

H. Graham Flegg
Reader in Mathematics, The Open University

Boolean Algebra provides a general introduction from first principles to the algebra of two—state devices through a discussion of sets, propositions and simple switching circuits. The algebra presented here now forms part of most 'modern' mathematics syllabuses and is of considerable importance in a number of fields of application.

POINTS AND ARROWS: 80p
THE THEORY OF GRAPHS
0 552 40003 3   160 pages

Arnold Kaufmann
Professor at L'Institut polytechnique de Grenoble.

Translated by H. Graham Flegg

Points and Arrows provides a sound elementary introduction to the theory of graphs, a branch of modern mathematics of increasing importance in various sciences, sociology, economics and business studies. Applications to various optimal path problems are discussed, and a fascinating glimpse is provided into recent theories of pattern recognition systems.

CALCULUS VIA NUMERICAL ANALYSIS   70p
0 552 40008 4   96 pages

A. Graham and G. Read
Senior Lecturers in Mathematics, The Open University

A novel approach to Calculus using the simple ideas of finite mathematics to introduce and illuminate more sophisticated notions. The common treatment of the two intimately related subjects of Numerical Analysis and Calculus is ideal for students with some brief knowledge of elementary Calculus, but no prior knowledge of Numerical Analysis is required.

RELIABILITY: A Mathematical Approach   70p
0 552 40009 2   96 pages

Arnold Kaufmann
Professor at L'Institut polytechnique de Grenoble.
Translated by Dr. A. Graham

Engineers, biologists, actuaries and many others are concerned with mortality (or survival) curves. For any system, whether technological or biological, the shape of its mortality curve depends upon its *reliability*.
This book offers a simple introduction to the mathematical basis of reliability and its associated concepts based upon a number of general examples.

These books are obtainable from your local bookseller. If you have difficulty obtaining them, you can buy them by post from the following address:

TRANSWORLD PUBLISHERS LTD.,
P.O. Box 11, Falmouth, Cornwall.

Please send with your order a cheque or postal order (not currency) to cover the cost of the book, plus 7p for each book to cover the cost of postage and packing.